Insulating Concrete Forms Construction

Demand, Evaluation, and Technical Practice

Ivan S. Panushev

Pieter A. VanderWerf, Ph.D.

Boston, Massachusetts   Burr Ridge, Illinois
Dubuque, Iowa   Madison, Wisconsin   New York, New York
San Francisco, California   St. Louis, Missouri

The McGraw·Hill Companies

CIP Data is on file with the Library of Congress

Copyright © 2004 by The McGraw-Hill Companies, Inc. All rights reserved. Printed in the United States of America. Except as permitted under the United States Copyright Act of 1976, no part of this publication may be reproduced or distributed in any form or by any means, or stored in a data base or retrieval system, without the prior written permission of the publisher.

3 4 5 6 7 8 9   BKM BKM   0 9 8 7

ISBN-13: 978-0-07-143057-9
ISBN-10: 0-07-143057-1

The sponsoring editor for this book was Larry S. Hager and the production supervisor was Sherri Souffrance. It was set in Century Schoolbook by International Typesetting and Composition. The art director for the cover was Margaret Webster-Shapiro.

McGraw-Hill books are available at special quantity discounts to use as premiums and sales promotions, or for use in corporate training programs. For more information, please write to the Director of Special Sales, McGraw-Hill Professional, Two Penn Plaza, New York, NY 10121-2298. Or contact your local bookstore.

Contents

Preface and Acknowledgments

We wrote this book to introduce contractors to the emerging building technology called insulating concrete forms (ICFs). ICFs are fairly new as construction products go. But because of their rapidly growing popularity, more and more contractors want or need to learn about them. For many, ICFs present an opportunity to expand their business by providing the customer with a superior product. For others they are a threat—competitors who use ICFs are attracting buyers and taking jobs that would otherwise have gone to a traditional contractor.

In this book we give you a basic understanding of ICFs, and we explain the things you need to consider when you're deciding whether to make the transition to building with them. For those of you who choose to adopt them, we tell how to get trained, choose a supplier, manage projects, and market your business.

We got our knowledge of construction and ICFs from a wide range of experience. Between us we have worked in different trades in the construction industry and as a general contractor. Later we founded a company called Building Works, Inc., which is dedicated to researching new construction technologies. At Building Works we have studied ICFs since 1993. In the course of that work we wrote numerous books, engineering reports, field evaluations, and construction manuals. We also worked on job sites with leading ICF contractors, went through most of the major ICF training courses, and trained other contractors ourselves. Currently we work on a variety of research projects for the construction industry. This keeps us at job sites and in constant contact with homeowners, builders, engineers, architects, and code officials.

For this book we have drawn from the experiences of over fifty ICF contractors from all parts of the United States and Canada. They range from small builders of single-family houses to large commercial developers. The factual information about ICFs comes from dozens of studies generated and supported by numerous private and public organizations such as the National Association of Home Builders, the U.S. Department of Housing and Urban Development, the Insulating Concrete Form Association, and the Portland Cement Association.

To make sure we provide you with accurate and useful information, we requested help and advice from a large group of people. We would like to extend special appreciation to the reviewers who pored through each draft to give us direction, add insights, and correct errors:

Buddy Hughes, *Insulated Concrete, Inc.*

Joe Lyman, *Insulating Concrete Form Association*

Will Oliver, *WMWN, Inc.*

David Shepherd, *Portland Cement Association*

We would also like to thank the following individuals and organizations who provided us with valuable insights, information, and materials about ICFs, without which this book would have been impossible:

American ConForm Industries

American Polysteel Forms

Bruce Anderson, *Vinyl Technologies, Inc.*

Phil Armstrong, *ECO-Block, LLC*

Arxx Building Products

Joe Ayala, *Ayala Construction*

Michael Bertin, *A.I.A.*

Karen Bexten, *Tadros Associates*

Ed Bobich, *American ConForm Industries*

Sunny K. Bresett, *Cellox Corporation*

Steve Brown, *Logix Insulated Concrete Forms, Ltd.*

Mark Buschman, *Buschman Homes*

Canam Steel Corporation

Joe Casassa, *Huntsman Chemical Corp.*

Tim Daughtry, *TriStar Development*

Dryvit Systems, Inc.

Tommy DuBose, *DuBose Construction*

Jim Eggeret, *Eggert Construction LLC*

Federal National Mortgage Association

Wayne Fenton, *LITE-FORM International*

FORMTECH International Corp.

Brian Fraser, *Fraser Construction*

Ian Giesler, *ICF Builders*

Ray Glasser, *R. Glasser Construction*

Lyle Hamilton, *Beaver Plastics*

Lambert van Haren, *van Haren Construction*

Charles Hoff, *Mansfield Township*

Duane Holloway, *Reward Wall Systems*

Lon Huff, *NW Reward Wall Systems*

ICF Accessories

Insulating Concrete Form Association

INSUL-DECK

International Building Code Council

Peter Juen, *Southeast Florida PolySteel*

Alex Kastenholz, *Elite Custom Builders*

Stanley Kerpoe, *Kerpoe Development*

David Lawson, *United Brotherhood of Carpenters and Joiners of America*

LITE-FORM International

Logix Insulated Concrete Forms, Ltd.

Jose Lopez, *Jose Lopez Builders*

Steve Lowrie, *Phoenix Systems & Components*

Randy Luther, *Centex Corporation*

Makita U.S.A. Inc.

Paul Mattaliano, *Mattaliano Contracting*

Vay Mickelson, *Mickelson Contracting*

Jim Miller, *Miller Brothers*

Keith Munsell, *Boston University*

Mark Napier, *INSUL-DECK Corporation*

Shane Nickel, *ReechCraft, Inc.*

Linda Norris, *Hambro Structural Systems*

Wayne Oke, *Oke Woodsmith Building Systems, Inc.*

Will Oliver, *WMWN Inc.*

John Passe, *U.S. Department of Energy*

Tom Patton, *Arxx Building Products*

Joe Peters, *Supreme Systems*

PHIL-INSUL Corp. O/A Integraspec

Joseph Pitsch, *Dan Vos Construction Co., Inc.*

Portland Cement Association

Jeff Preble, *Standard ICF*

QUAD-LOCK Building Systems Ltd.

Brad Reed, *City of Lubbock, Texas*

Reward Wall Systems

Jerry Rhodes, *Hambro Structural Systems*

Cameron Ridsdale, *F.A.I.R.C.*

Richard Rue, *Energy Wise Systems*

Edward Scherrer, *Polysteel Supply of Minneapolis/St. Paul*

Alan Schofield, *Schofield Concrete*

Dean Seibert, *Wind-Lok Select*

Scott Sinner, *Scott Sinner Consulting*

Southeast Florida Polysteel

Gerald Spude, *Wisconsin Thermo-Form, Inc.*

Andy Stephens, *Livingston's Concrete*

Cooper Stewart, *Logix Insulated Concrete Forms, Ltd.*

Kurt Thompson, *KDT Construction, Inc.*

U.S. Census Bureau

U.S. Green Building Council

United Brotherhood of Carpenters and Joiners of America

Ray Valdes, *Rastra USA*

David Watson, *American Polysteel*

Randy Wilcox, *Randy Wilcox Contracting*

Kelly Willoughby, *Kaw Valley Chapter of Habitat for Humanity*

Wisconsin Thermo-Form, Inc.

Danny Yochum, *ICF Accessories*

And finally, we would like to thank the editors at McGraw-Hill and ITC who spent countless hours refining and improving on our manuscript:

Larry Hager

Margaret Webster-Shapiro

Waseem Andrabi

To our readers, it's obvious that you're interested in constructing better buildings. We wish you success.

Ivan S. Panushev
Pieter A. VanderWerf, Ph.D.

ABOUT THE AUTHORS

IVAN S. PANUSHEV is a licensed engineer who has researched ICFs for the past three years. He is a graduate of three ICF training programs and labored on several job sites to learn the details of ICF construction first-hand. His articles on ICFs have appeared in such magazines as *Concrete Construction*, and he is the author of the *Prescriptive Method for Connecting Cold-Formed Steel Framing to Insulating Concrete Form Walls in Residential Construction*.

PIETER A. VANDERWERF, PH.D., is an expert on insulating concrete forms in construction who has been researching the subject since 1993. A columnist for *Permanent Buildings and Foundations*, a trade industry publication, he is the co-author of two other books on insulating concrete forms: *Insulating Concrete Forms for Residential Design and Construction* and the *Insulating Concrete Forms Construction Manual*. He also co-authored *The Portland Cement Association's Guide to Concrete Homebuilding Systems*. (All three books are published by McGraw-Hill.) He has served as general contractor on ICF homes and for a year he moderated an online ICF message board.

Introduction

Do you build houses or low-rise commercial buildings? Would you be interested in a new, superior way to build the structure? One that is flexible, superinsulated, and has a core of steel-reinforced concrete? One that is selling fast and growing fast? One that buyers willingly pay a premium for over wood and steel frame?

This book is for people who answer "Yes" to all those questions. It is about insulating concrete forms (ICFs) and how to start building with them.

ICFs are hollow foam forms that are used to create concrete walls and even floors and roofs quickly and effectively. The end result is a *sandwich* wall with a layer of construction-grade foam on each side and steel-reinforced concrete in the middle.

ICF walls are superinsulated and structurally superior to other alternatives for low-rise buildings. In homebuilding they can cut a month off the construction schedule and give the contractor the opportunity to build the whole house from the sub-grade to the roof. In commercial construction, using ICFs can shorten the construction cycle and reduce the amount of coordination required between trades. But above all, ICFs produce a better building that a growing number of buyers are looking for and willing to pay a premium for.

ICFs are not for everyone. They require training and an open mind. They cost a bit more than frame, so you have to pick your markets carefully and learn how to present the benefits to the customer. But with commitment and the right plan you can immediately become more competitive in the market place because you offer a premium product at a reasonable price. If you are happy with your existing business and do not want to diversify or expand, this book is probably not for you. But if you are excited about the possibility of providing a better end product to your customers and getting more business because you do, you have in your hands the best piece of information available to learn about how to do it.

ICFs were invented in the 1940s in Europe. They made big progress after modern plastic foams and molding technologies were developed in the 1960s. They became widely commercially available in North America in the 1980s. But even then there were only a handful of companies in Canada and the United States experimenting with this radical method of forming concrete using foam that was left in place. With still more improvements and a lot of hard work by

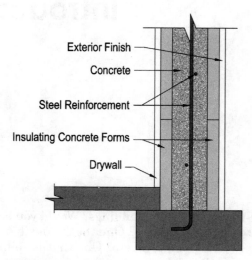

Figure I-1   Cross section of an ICF wall.

a lot of contractors who streamlined installation, things began to take off in the mid-1990s. Based on data from the National Association of Home Builders, 48,000 houses were built with ICFs during 2002. An estimated 10,000–15,000 low-rise commercial buildings were also constructed. Sales have been growing at the rate of about 30 percent annually, and show every sign of continuing to grow rapidly in the future.

Who and What the Book Is For

This book is a guide for carpenters and general contractors considering getting into the business of building with ICFs in order to get an edge over their competitors by offering a better product.

To keep the book objective, the information in it comes mostly from other contractors who have been in the same position. There are stories from the builders who switched their businesses to ICF construction and the owners who bought the buildings they built. These experiences tell us what things worked well, what ideas turned out to be mistakes you should avoid, and how you can appeal to the buyer.

Warning.   This book is written for the *experienced contractor*. It assumes that you know the basics of frame construction and running a contracting business. In this book you will *not* learn how to be a builder, but how to make the switch from building with sticks, fiberglass, and plywood to building with foam, steel, and concrete. The book tells you what new things you have to do, or what old things you have to do differently, to switch your business to ICFs and do it profitably.

Why a Contractor Might Be Interested in Building with ICFs

Building better buildings is admirable, but contracting is a business and it has to make money and provide financial support for its employees and owners. For many contractors, ICFs can provide an opportunity to do all these things at once. The many advantages of the product can help you attract buyers away from conventional contractors. By directing your business to the right buyers in the right way, you can also command a premium price that gives you an attractive profit.

Over a decade of rapid growth proves that ICFs are here to stay. Now you can ride that growth to build your business too. On the other hand, if you stick with old methods of construction, you run the risk of losing more and more sales to others who have made the switch.

One industry analyst predicted that the development of the ICF market will be similar to that of engineered wood joists. They appeared about 30 years ago. They were a little more expensive and unfamiliar to the typical builder. But they have grown steadily and now hold a solid 45 percent of the joist market. So if you are a carpenter or a general contractor and do not want to be left behind this wave of change you should consider ICFs. Plenty of opportunity exists for new builders to enter the market, and this book will teach you what steps to take to do it.

What Does the Book Include?

In order to learn how to build a successful business you should first understand *exactly what insulating concrete forms are and can do*. You will find valuable information here on their properties and why consumers are interested in them.

One chapter is devoted to the *construction process* of ICF walls. It gives enough detail so you understand just how different ICFs are from frame, without weighing you down with a lot of detail that is unnecessary at this point. When you choose which ICF to adopt, most manufacturers provide excellent training programs and product literature. These will tell you about the nuts and bolts of the installation process in great depth. This book is not a textbook or a construction manual, but a business guide describing all the opportunities that this new technology provides and how best to utilize them.

Later you will learn how to *choose an ICF supplier* and how to evaluate each supplier's product and services. You will find information on the best way to *get installation training*, since this is one of the most important steps in your move to insulating concrete forms. There are also discussions on how to *find or assemble a good ICF crew*, how to *coordinate between contractors*, and some tips from other contractor converts on how to *keep business rolling* once you have switched.

The book also covers the various *grants, awards, and loans* available to builders and owners of ICF homes, based on energy efficiency or superior construction. This includes what they are, how they work, and how to get them. You will see suggestions on how to deal with building officials, engineers, and architects.

The book also covers special considerations in getting into commercial ICF construction.

Finally, there is plenty of information on the *marketing techniques* that will help you find customers and make a successful sale. At the end you will find a discussion on the *future of ICF construction*, its expected growth, possible new products on the horizon, and new programs that may be enacted in the near future to stimulate ICF and energy-efficient construction.

Other ICF Resources

The book refers to a lot of other information sources. Some of these have so much useful information that it's worth highlighting them right now.

Associations

A great source of help and information is the *Insulating Concrete Form Association*. Eventually you should consider joining this organization. It offers plenty of materials on ICFs, building codes, and engineering. It also supplies marketing materials and runs educational seminars for ICF contractors. In the near term you can check out their web site (*www.forms.org*). The site includes listings of ICF manufacturers, distributors, accessory suppliers, and contractors. As with any good association, you will get out of it what you put into it. If you join and become involved, you will find yourself making important contacts in the industry and keeping a step ahead of others on important news.

Two other organizations that provide information and support are the *Portland Cement Association* (*www.portcement.org*) and the *Cement Association of Canada* (*www.cement.ca*). They have created information materials to help you get trained and market ICF construction. The Cement Association of Canada has a training course specifically on ICF installation and the Portland Cement Association has classes and seminars on concrete use in residential construction. During the last five International Builders Shows the Portland Cement Association erected ICF show homes for public viewing. It sometimes also has show homes and other major displays on ICFs at the annual World of Concrete Show. As you'll see when you read the book, a lot of useful books and reports and flyers about ICFs are available from these associations. You can order them at their web sites or by calling the PCA Publications toll-free line (800-868-6733).

Shows

Consider visiting the International Builders Show (*www.buildersshow.com*) and the World of Concrete (*www.woc.com*) yourself to meet the vendors of ICFs and see their products first-hand. It's a great opportunity to gather a lot of information and meet a lot of helpful people in one place.

Books

There are also several useful books:

- *Insulating Concrete Forms for Residential Design and Construction* by Pieter A. VanderWerf, Stephen J. Feige, Paula Chammas, and Lionel A. Lemay (McGraw-Hill, 1997) gives detailed technical information on ICFs, including products, physical properties, design, engineering, and construction techniques. It is useful for architects, engineers, and technically-oriented contractors.

- *Insulating Concrete Forms Construction Manual* by Pieter A. VanderWerf and W. Keith Munsell (McGraw-Hill, 1995) is an earlier book which provides an overview of common ICF construction methods.

- *Prescriptive Method for Insulating Concrete Forms in Residential Construction* by the National Association of Home Builders Research Center (2002) is a comprehensive engineering manual for the structural design of the ICF walls of houses. Its contents have been adopted by major model building codes, which now accept ICF construction. The Second Edition of the book appeared recently. It includes the additional information needed for ICF engineering in high-seismic regions, and is available from the Portland Cement Association.

Web sites

The Internet also provides accessible information on ICFs. One of the most useful sites is *www.ICFweb.com*. This independent site features an open discussion forum that lets you post questions for builders, engineers, and architects. Some of the comments are pretty opinionated, but it can be useful to hear what others in the business have to say. There you can also search for ICF professionals in your area and access information on projects that are open for bidding.

Some other sites worth visiting are the ones of the ICF manufacturers. You can access these through the "Member Search" feature of the Insulating Concrete Form Association's site at *www.forms.org*.

The ICFA site is itself also very useful. The "Member Search" connects you to a directory that shows not only the major ICF manufacturers, the directory also includes names of ICF contractors and distributors, and suppliers of related products.

Another very useful site is *www.concretehomes.com*. Operated by the Portland Cement Association, this site provides clear, concise descriptions of alternative construction technologies and listings of other available resources. Most of the key ICF books and reports and brochures are available for sale on this site.

Other useful sites include the following:

- *www.huduser.org*—Department of Housing and Urban Development (HUD). This includes many government documents and reports.

- *www.energystar.gov*—Energy Star Homes. This provides information on the Energy Star Program for energy-efficient homes.

- *www.nahb.org*—National Association of Homebuilders (NAHB). This includes useful marketing information.

- *www.nahbrc.org*—National Association of Homebuilders Research Center (NAHBRC). This site includes several field studies of insulating concrete forms.
- *www.pathnet.org*—Partnership for Advancing Technology in Housing (PATH). This site provides information and reports about the U.S. government's PATH program for new building technologies, including information about ICFs.

Magazines and newsletters

There are two magazines that regularly have articles and columns covering insulating concrete forms.

- *Concrete Homes Magazine* (*www.concretehomesmagazine.com*) includes news and feature articles about all types of homes built with concrete, including ICFs.
- *Permanent Building and Foundations Magazine* (*www.pbf.org*) covers different concrete forming technologies, but it has a strong emphasis on concrete homes and especially homebuilding with ICFs.

You will also find occasional articles on ICFs in *Concrete Producer* and *Concrete Construction*. You can get information about these at *www.hanley-wood.com*. Click on "Products" and find the names of these magazines.

The *ICFA Newsletter* is also a very useful source. It not only has news about developments, it has news about some developments *before* they happen. Since it is the newsletter of the Insulating Concrete Form Association, it alerts readers to upcoming projects as well as current industry events and activities. You get it for free when you join ICFA.

The Portland Cement Association publishes a newsletter called *Concrete Homes*. Don't confuse this with the magazine of the same name. It also covers developments in the business of constructing homes out of all types of concrete walls, but in a briefer, monthly newsletter format. It also alerts readers to new information and promotional materials that become available. Go to *www.concretehomes.com* and click on "Newsletter" to subscribe or view back issues on-line. It's free.

How to Use This Book

It's time to start deciding whether ICFs are for you and learn how to adopt them. There is no right or wrong way to reading the material in this book. The chapters are laid out in a logical order, but feel free to read them any way you want. Just make sure you go over everything before you make any really big decisions.

Good luck!

Advantages of ICF Buildings

Overview

The real reason for interest in insulating concrete forms (ICFs) lies in the benefits they provide to the owners and occupants of the building. They're the reason so many buyers want ICFs, and that's why you can increase sales and profits building with them.

It's critical that any contractor offering ICFs understands these benefits. Buyers ask lots of questions. Some of them get very detailed. To sell the product effectively you have to be able to explain what the buyers get out of it, and you have to know what you're talking about or they'll be skeptical.

Understanding the benefits is also important for building with ICFs. It helps you appreciate why things are done the way they are, and what the harm might be from using shortcuts. That helps guide your work. Alan, a contractor in the Northeast, tells about an experience on a custom home he built:

> This was my first ICF project. We had help, but we made some mistakes. We had one sliding door opening, so we had to form a long lintel [header] over it. I knew that if we formed it with ordinary plywood and not the foam, the concrete would be wider and we could get the strength for the span. So I started to plan it out this way. Then it hit me—this would leave a big uninsulated spot in the wall. Half the point of using ICFs was that it was great insulation, and that's what the owner was paying for. So I asked our ICF supplier how to get enough strength with their regular concrete thickness inside the block. It turns out we just needed to add some more rebar. The foam was still there, the wall was still well insulated, and it was actually easier doing it that way without building plywood forms.

After you go through the description of benefits in this chapter you should be able to explain them to others and make the right decisions so that you provide all those benefits too.

Most ICF buildings would have been built out of wood frame otherwise, and that's usually what the buyer is comparing the ICFs to. So in discussing the benefits of ICFs this chapter usually compares to frame, too.

Energy Efficiency

Energy savings is one of the most important features of ICF buildings to buyers. But it's also the most complicated benefit to estimate, measure, or explain. So there is a lot of information here about energy efficiency. Don't be intimidated. You should learn it all some time, but you can skip it or skim it now and come back to it again later.

So why is it so hard to just tell someone "If you build your exterior walls out of ICFs, then your fuel bill will be X percent lower"?

Simply put, Your Mileage May Vary. How much energy you save depends on dozens of different things. A *partial* list of important factors is as follows:

- The local climate
- How much window area the building has and how energy-efficient the windows are
- How well the roof is insulated
- The efficiency of the heating and cooling equipment
- How tight the roof construction is
- How good the construction is that you're comparing to
- How many people occupy the building and how much they go in and out
- The set point the thermostat is kept at

These things make a big difference even when they are kept the same in the ICF building and building you're comparing it to. They make a *huge* difference if they're different for the ICF and the other building because then you're no longer comparing apples-to-apples.

Studies and stories have produced estimates for overall energy savings of anywhere from 12 to 75 percent. But there is some reason to be skeptical about either of the extremes.

Some of the highest savings numbers come from owners of ICF houses who compare their fuel bills with the bills of their neighbors. When the houses are of similar size and the rest of the construction besides the walls is about the same, these might be valid comparisons. But sometimes other things are very different, too. Jack, a homeowner from Nebraska, tells a questionable story about his house:

> My house is built of ICFs, and it uses 70 percent less gas for heat than my neighbor's house. His is regular wood frame, and it's about the same size. He let me have some of his utility bills to compare, so I can prove it.

But Jack's house also has sixteen inches of blown-in cellulose insulation in the attic, double-pane low-E argon-filled windows all around, a ground-source heat pump, and probably a bunch of other energy-saving measures as well. There's no way to say that all of his savings came from using the ICF walls. Some, yes. All, no.

On the flip side, some of the lowest estimates came from "scientific" engineering studies that use computer models, or run lab tests on actual wall assemblies, or run tests on specially-constructed houses to estimate the energy consumption in same-design ICF and frame buildings. This sounds great, but the frame walls are always constructed (or modeled on the computer) with near-perfect details. Richard, an energy consultant in Texas, talks about how inaccurate this is. He recalls some of his very first field research on frame houses:

> We went out and inspected some houses before the drywall went on, and I was blown away. In half the bays the insulation didn't go down to the bottom. There were gaps of 6 inches in some of them. Those were totally uninsulated areas. And the insulation was stapled to the sides of the studs. It's supposed to be stapled on the face of the studs so you don't compress it. But they just went ahead and compressed it. I learned later that a lot of insulation contractors feel like they can't staple on the front of the studs because the drywallers don't like it. The uneven paper and the staples interfered with putting the drywall on. We kept visiting sites and I kept on seeing this stuff, right to this day. It was a few years later I saw ICFs. If you screw up the insulation on frame walls you get away with it. If you screw it up on ICFs the wall blows out on you, so you damn well make sure you get it right.

So in most cases the actual savings are probably going to be somewhere in between.

What you can say

Go ahead and swap stories about the savings that some local homeowner got out of his new ICF house. Your local ICF distributor will probably have a few. You'll hear more from other contractors and from owners. Just make sure everyone realizes that these examples are not a guarantee.

There is also a different message you can give buyers, that *is* effective. It goes like this: "If you build your walls out of ICFs, you have effectively eliminated the walls as an energy concern. You'll get some good savings. Your total savings will then depend on how well you handle the other weak spots—roof, windows, and heating and cooling equipment."

To put it another way, using ICFs does about all you can do to make the walls energy-efficient. If you are as careful with the rest of the building, you'll have an extremely economical building to heat and cool.

You can also explain to people *why* ICFs are about as energy-efficient a wall as you can get. The reason is that they save on the heating/cooling in all three of the most important ways:

1. They have a high R-value.
2. They are very air-tight.
3. They have high thermal mass.

If you can explain each of these a little you will impress the customer all the more.

R-value

R-value measures how fast or slowly something conducts heat. **Conduction** is the scientific term for what happens when something hot warms up the things it is touching, like when hot tea poured into a cup heats up the cup.

Conduction accounts for a lot of the energy loss from a building. In winter, the warm air inside hits the walls (and windows and doors and ceiling) and warms them up, and the warm walls touch the cold air outside and warm it up. In this way the inside air gradually loses its heat, it gets cool indoors, and the furnace has to come on to get the temperature back up. The same thing happens in reverse in the summer, when warmth gradually comes indoors and the air conditioning kicks on.

Some materials slow down this conduction process better than others. The best ones we call insulation, and we put them in the walls to keep energy bills low. We measure how well a material or a wall slows down the conduction with a number called the **R-value**.

The formula for R-value is something every mechanical engineer learns. For our purposes it's enough to know that the higher the R-value of an insulation or wall, the better an insulator it is and the slower heat will conduct through it. For example, if I replace a wall that has an R-value of 10 with another one that is R-20, heat conducts through the new wall half as fast. So the heating/ cooling has to run less, and less fuel is necessary to keep the indoors at the desired temperature (see Fig. 1-1).

Most of the wood frame homes in North America are built with fiberglass insulation in the walls that is either R-11 or R-13. The walls as a whole don't have quite as high an R-value, however, because only about 75 percent of the wall area has insulation in it. The rest is solid wood—studs, plates, headers, and so on. Laboratories that have put well-constructed wood frame walls up on their testing

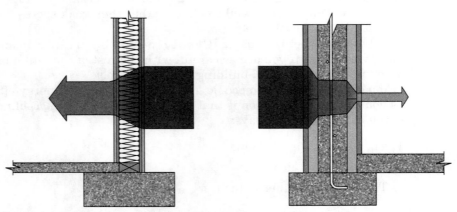

Figure 1-1 Because ICFs have a higher R-value, heat conducts faster through a frame wall (left) than through an ICF wall (right).

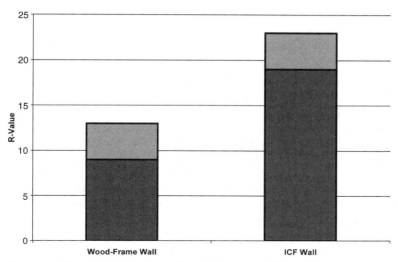

Figure 1-2   The range of R-values of typical frame and ICF walls.

racks have found that their actual R-values have usually measured in at about R-8 to R-12.

ICFs , in contrast, have two foam layers that provide an R-value of anywhere from about 16 to 23. The reason for the variation is that different ICFs use different thicknesses and types of foam. The concrete inside the foam adds a hair to the R-value, so the total wall R ends up around 17 to 24. And there are very few breaks in the insulation, like the ones that reduce the R-value of a frame wall.

The bottom line is that a typical ICF wall has around twice the R-value of the typical wood frame wall. So you would expect an ICF wall to lose about half as much heat from conduction as the frame wall (see Fig. 1-2).

Air-tightness

Infiltration is the leaking of outside air into the building, and it is a bigger source of energy loss than most people think. When cold air leaks inside during winter, or hot air leaks inside during summer, the heating/air conditioning equipment has to run to get the indoor temperature back to the set point. ASHRAE (American Society of Heating, Refrigerating, and Air Conditioning Engineers) estimates in its handbook that air infiltration accounts for 20 to 40 percent of the energy used for space conditioning in houses in the United States.

Frame walls are assembled from thousands of pieces that are mechanically fastened to one another. Of course there will be tiny gaps between them at many points. According to Larry, a homebuyer in Michigan:

> This weekend I went down the road where a stick frame house is in the same phase of construction as mine. I took my trusty feather with me to check around the

windows, doors and electrical boxes. On every window and door and about 50% of the boxes I could detect a draft! I think the quality of ICFs vs. stick frame is quite obvious here. I am still living in an apartment, but I can't wait to get into my new ICF home!

ICF walls have a big advantage over frame construction when it comes to air infiltration—the concrete cast into the forms seals the wall (see Fig. 1-3). How big the difference in air infiltration is between frame and ICF buildings is one of those cases where your mileage will vary. The amount of air passing through the ICF wall is probably about as near zero as a wall can get, but how tight the entire building is also depends on how well other things are sealed.

But tests of actual houses have shown that there is a definite improvement with ICFs. A study of a lot of new frame homes published in the ASHRAE handbook showed that they averaged about 0.5 **air changes per hour** (ACH). ACH is a common measure of the level of air infiltration. Some ICF houses have been measured at an ACH as low as 0.15. But more typically they come in at around 0.25 or 0.33 ACH. So that suggests the air infiltration can be cut somewhere between a half and a third, compared to typical frame construction (see Fig. 1-4).

Thermal mass

Thermal mass is the ability of heavy materials to store heat. When your building shell has a lot of thermal mass it can even out the extreme temperatures outdoors. This helps to save even a little more energy.

A clear example of thermal mass comes from the adobe homes built by the Native Americans of the Southwest. Southwestern desert temperatures can reach 100°F during the day and 40°F at night. But the heavy adobe walls help keep things inside tolerable. During the day the sun strikes the walls and heats them up. But they are still cool from the night before, so they help keep the indoors cool. And because of their mass, they only warm up, very slowly.

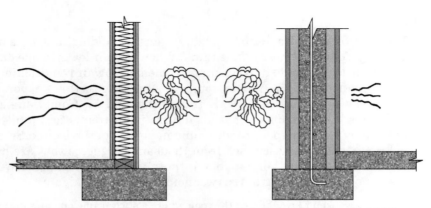

Figure 1-3 More air infiltrates through frame walls (left) than ICF walls (right).

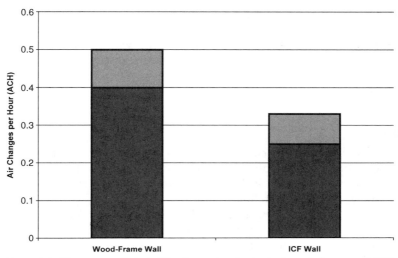

Figure 1-4 The range of air infiltration rates for typical wood frame and ICF buildings.

By nightfall the walls may be quite warm. But then the sun is gone and the night air is cool. Instead of being a problem, the warm walls actually help to keep the indoors warm at night. Then the next day the cycle starts all over again. In short, the thermal mass of the walls buffers the interior from the temperature extremes going on outdoors (see Fig. 1-5).

ICF walls have a lot of thermal mass because of the concrete in the core. This helps save some more energy, but estimates on how much differ. One big reason

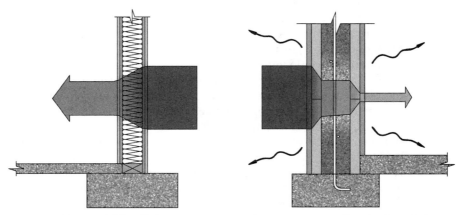

Figure 1-5 Because ICF walls have greater mass, heat moves more directly through frame walls (left), while it gets more absorbed and slowly released from ICF walls (right).

is that the savings depend a lot on the local climate. Thermal mass has the biggest effect in temperate climates, where the outdoor temperature hovers around 70°F or so for a major part of the year. But the engineers say that even in consistently cold places like Minnesota and consistently hot places like South Florida, the mass should be good for a few percent in added energy savings.

Equivalent R-value

This is one of the most confusing things about energy efficiency, but you're going to hear about it and you'll probably have to explain it to buyers at some point. So you'll have to learn it. You can do it now or skip this section for a week and come back when you're feeling up to it.

Here's the basic story of **equivalent R-value**. Suppose I have a frame house and an ICF house of the same design. The ICF house uses less energy for heating and cooling. The owner of the frame house might say, "That's because its walls have a higher R-value. They're about R-20, and mine are only R-10." So he might come up with a plan to get the same level of savings as his neighbor. He might tear down his house (OK, this is only a story) and rebuild it with thicker walls that have more insulation. And he might make sure the work is done well. And by the time he's done he might truly and honestly have frame walls that are R-20, the same as the ICF walls.

So, will he now have the same energy bills as his neighbor in the ICF house? Well, no, actually he won't. The higher R-value walls should cut conduction to be the same, but the ICF walls still get some additional savings because of their lower air infiltration and their high thermal mass. So now someone selling ICFs says to himself, "You know, when people go out shopping for houses and they compare energy efficiency, they always look at R-value and that's about all they look at. But that's not fair. Our walls rate about twice the R-value of frame walls, but even if the frame walls were built so that you doubled their R-value, our houses would still use less energy because we have lower air infiltration and higher thermal mass. So when they look at R-value they only get part of the picture, and I can't get them to look at anything else."

But the ICF salesman could have a bright idea. He could say to himself, "You know, I wonder how high you'd have to jack up the R-value of the frame house to get down to the same energy bill as the ICF house?" Then he could pay the owner to rebuild the frame house *again*, making the walls thicker and adding more insulation until he got the fuel bill down to the exact same amount as the ICF house. And when he finishes he could look at the walls and see that they had to be, say, R-40 to give that frame house the same level of savings as the ICF house. Because the frame house never gets the same savings from lower air infiltration and from thermal mass, you have to keep stuffing insulation into the walls to try to compensate.

Now the ICF salesman goes out and tells people, "An ICF wall has an *equivalent* R-value of R-40. By that I mean, to get the same energy bill as an ICF house you would have to build your frame walls so that they were R-40. So building with ICF walls is *equivalent* to building with R-40 frame walls."

Now the ICF salesman has an R-value number to hand out that gives people some kind of idea of the total energy savings from ICF walls. Equivalent R-values this high are not unusual. A few years ago, engineers at Construction Technology Laboratories did energy consumption estimates for ICF and frame houses that showed that a house with ICF walls would *still* have a lower energy bill than even an R-38 frame wall constructed with 2 × 12 studs. So that means that their equivalent R-value would be somewhere over 38. And this result holds for different climates across the United States. The report of this study is titled *Energy Use of Single-Family Houses with Various Exterior Wall Systems*. Some people refer to it, and if you want a copy you can order it on the *www.concretehomes.com* web site or at PCA Publications (800-868-6733).

A lot of people find equivalent R-value to be a useful way to summarize the energy efficiency of ICFs. And whether you do or not, other people will use it so you might as well know what it is. But there are a few important cautions.

First, equivalent R-value is not the same as the conventional R-value and you shouldn't pretend it is. ICF walls are *not* R-40. They still allow heat to conduct through like R-20 walls because that's what they are. It's just that they have other, different ways to save some energy. Equivalent R-value is a way to boil all the energy savings down and summarize them in one number. But it's a different type of number.

Second, there's no good way to figure out total fuel bills from the equivalent R-value. If a frame wall has an R-value of 10 and an ICF wall has an equivalent R-value of 40, that does *not* mean that the ICF building will have one-quarter of the fuel bill. Without an engineering degree, all you can really say is that the higher the equivalent R-value, the greater the energy savings.

Third, the equivalent R-value depends as much on location as it does on the walls. So a certain ICF wall doesn't really *have* a set equivalent R-value. The equivalent R-value of the same wall could be 50 in St. Louis and 40 in Miami and 30 in Edmonton. The thermal mass savings are lower in extreme climates, and so the equivalent R-value is lower.

Fourth, just to confuse things even more, some people use a different term than "equivalent R-value." Some call it **effective R-value** or **mass-corrected R-value**, but they all mean about the same thing.

So use equivalent R-values carefully. They're kind of neat, and they can get across the idea that there is more to energy efficiency than what the conventional R-value number tells you. Just don't pretend that they tell exactly how much energy you can save in a particular building.

Wind Resistance

Of all natural disasters, high winds from hurricanes and tornadoes cause, by far, the greatest loss of life and property in the North America. Fortunately, ICFs offer good wind resistance.

Structural endurance

During the 1990s, two separate teams of researchers surveyed the destruction from Hurricane Andrew in Florida, and one visited the site of Hurricane Iniki in Hawaii. This was useful because both areas have a lot of small buildings built of conventional frame and a lot of others built of reinforced concrete. Most of the concrete buildings were reinforced concrete block, but structurally reinforced block walls behave about the same as other reinforced concrete walls, like ICFs.

All the research teams reached similar conclusions. A significant number of the buildings lost their roofs. The reports didn't give exact numbers, but it appears that in some spots roofs came off of over 30 percent of the buildings. This takes an important part of the building structure away, and a lot of the frame buildings that lost their roofs also collapsed completely. But almost none of the reinforced concrete buildings did. Observers often saw twisted piles of wood studs surrounding a neat concrete box that was still intact.

This is about what you would expect. Concrete's weight and the hold-down power of the continuous lines of steel rebar from foundation to roofline give reinforced concrete buildings a large advantage against the uplift and side forces of wind.

Structurally, tornadoes can be even more devastating than hurricanes. Major hurricanes can have gusts of 250 mph. They can have sustained winds of 200 mph.

In one case that appeared in the news, an ICF house in Washington, Iowa was in the direct path of a tornado with winds up to 200 mph. It survived largely intact except for some damage to the roof and some siding that got ripped off. Yet half a dozen frame houses next to it were ripped off their foundations (see Fig. 1-6). The owner, Reverend Monte Asbury, was not at home at the time, but his children were. According to Asbury:

> When Hannah and Ben [the children] heard the sirens, they grabbed the essentials—pop and chips—and headed for a windowless room in the basement. They said it felt like their ears were being sucked out of their heads, so there must have been a tremendous change in the atmospheric pressure. The kids didn't even hear the tree that hit the house. The contractor who helped build the house said he was absolutely certain a wood wall would have been crushed by the tree that fell. Several people have come up to say "I'll bet you're glad you built a concrete house."

The Asbury house had extra strong tie-downs from the roof to the ICF walls, but not much else special to resist the wind.

If you want to build a house to resist even the strongest winds, ICF walls can be readily designed to meet the challenge. It usually requires a slightly thicker wall and more rebar. A concrete roof is also advisable.

A less-expensive alternative than an all-concrete house is the "safe room." A safe room is an interior room of a house, usually a walk-in closet, that is constructed to withstand a tornado. Occupants are supposed to run into the room for safety when there is danger. Several ICF companies regularly sell the materials to build ICF safe rooms (see Fig. 1-7). These are most popular in high-tornado areas like Oklahoma. Plans for ICF safe rooms are even included in publications from the U.S. Federal Emergency Management Agency.

Figure 1-6   The remaining concrete foundation from a demolished frame house (above) and the neighboring ICF house (below) in the direct path of a tornado. (*Courtesy of Reward Wall.*)

Resistance to projectiles

Some scientists report that most of the injury and damage from tornadoes and hurricanes actually results from flying debris. Fortunately, ICFs' combination of outside foam layers and a reinforced concrete core appears to be highly resistant to impact damage.

Figure 1-7 Preassembled ICF safe room ready to ship to the job site. (*Courtesy of Life-Form International.*)

In 1997, the Wind Engineering Research Center at Texas Tech University performed tests on frame and ICF walls to measure projectile penetration. They shot a wooden stud at the walls at speeds ranging up to 119 mph. This simulated the speed of debris swirling around the outside of a tornado. All the walls had standard wallboard on the inside. The frame walls had their standard plywood sheathing outside. Half of the frame walls and half the ICF walls had a standard brick veneer outside, and half of each had vinyl siding. The stud completely penetrated all of the frame walls, no matter what speed the stud was traveling or what outside finish the wall had. The stud never penetrated any of the ICF walls, and the researchers reported that they didn't even see any cracking. Some of the ICF walls were fired upon twice, still with no damage (see Fig. 1-8).

Figure 1-8 A stud shot at over 100 mph dents the siding and bounces off an ICF wall (left), but goes entirely through a frame wall from outside siding (middle) to interior wallboard (right). (*Courtesy of Portland Cement Association.*)

Fire Resistance

The massive, noncombustible concrete of the ICF wall gives it certain advantages in a fire. However, be careful *not* to assume that this allows occupants to relax their fire safety measures. There are still many fire hazards that no ICF wall—in fact no wall of any kind—can prevent.

The advantages of concrete walls are that they stand up well structurally in a fire, and they slow down the passage of fire from one side of the wall to the other. Both can be valuable, depending on the circumstances. But the risk of smoke inhalation remains, and it's about the same as with wood frame.

Consider the course of a typical building fire. When a fire starts indoors, it almost always burns many of the contents of the building before it ever gets through the wallboard and into the exterior walls. This creates a lot of smoke. In fact, most victims of fire die from inhaling this smoke and suffocating. There is nothing you can put in the exterior walls—no matter what they are made of—to prevent this. When the smoke starts to get bad the walls are not even involved yet. So all the usually required safety measures—having working smoke alarms, planned escape routes, and so on—still apply.

If a fire progresses far enough, it can penetrate the wallboard and get to the inner contents of the exterior walls. It can destroy either wood or foam, but not usually concrete. In a wood frame wall, the burning of the wood adds fuel to the fire, adds smoke to the air, and eventually can collapse the structure as.

ICF foams also create smoke, but they do not generally fuel the fire. The foams used in ICFs today are required to be what is called **modified**. This means they contain an additive that prevents them from supporting combustion. In other words, you can't light them. If you hold a flame up to the foam of an ICF, it will melt. But as soon as you move the flame away, the melting will stop and the ICF will not burn (see Fig. 1-9).

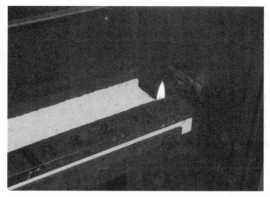

Figure 1-9   Modified foam melts away from a flame in a burning test. (*Courtesy of Huntsman Expandable Polymers Company, LLC.*)

But in a real fire the foam may be subjected to continued exposure to flames because the other building contents will be burning all around it. In this case the foam will give off smoke much like wood does. In an extensive study, the Southwest Research Institute concluded that the emissions from burning foam are no more toxic than the emissions from burning wood. That's nice to know, but it also serves as a warning that you need to take the same safety precautions in covering ICF walls as you do in covering frame walls. There is no justification for reducing the amount or fire resistance of the wallboard covering the wall. The wallboard is still necessary to slow any fire from getting to the inner part of the wall.

On the other hand, once fire gets to the inner wall, a concrete wall is much less likely to fail structurally. It takes uncommonly high temperatures—in the thousands of degrees—to weaken concrete significantly. Since these rarely occur in a building fire, ICF walls will probably stay standing no matter how long the building burns. This is some help if there are occupants who still have a chance of getting out. They will at least not be impeded by collapsing walls. It also may reduce total property damage.

In addition, it takes heat and fire longer to get through an ICF wall than a frame wall, and that is good news on a few fronts. If there are any interior walls built of ICFs, those walls will help contain the fire. That gives occupants on the other side more time to get out and the fire department more time to put out the fire before further damage. It is common to use ICF or other concrete walls as the dividing walls between apartments in multifamily buildings or between units in other large buildings, exactly for this reason.

In a case that hit the news in 1999, one of the units of an apartment complex in Beaufort, South Carolina had a fire inside. The fire was so intense that it melted the external light fixtures outside the unit. But the complex was built with ICF walls, including the walls between units. According to John, the property manager:

> The [ICF] walls performed so well that the walls of the adjoining unit did not burn and in fact were not even warm. And the only cleanup required for the other seven units in the building was to shampoo the carpets and clean the air ducts. Restoring the burned apartment was quick and easy because all the walls were still in tact.

Even if only the exterior walls are ICFs, they may slow the fire from getting outside the building and spreading to other buildings and outside objects. Or if the fire starts outdoors, the ICFs may slow the fire from coming inside. This can be important in hot, dry climates where brush fires are common. It can also be important if neighboring buildings catch fire.

According to Kurt, an ICF contractor in Utah:

> I built my own house out of ICFs. One time the house next door caught fire and it was a big blaze. It was right up against my house at points, but all that happened is that it melted away some of the outside foam. I'm going to repair it to replace the foam, but I don't really have to.

Figure 1-10 Hose steam on an ICF wall after 4 h of heat exposure. (*Courtesy of American Polysteel.*)

The ability of walls to stop the passage of heat and fire has been tested many times. The United States and Canada have similar standard procedures for the test. It begins with a gas flame or electric resistance heating directed at one side of a wall. The temperature of the heat is controlled, and is increased over time. The test stops as soon as any one of three things happens: the fire burns a hole in the wall, material on the opposite side of the wall catches flame, or the temperature on the opposite side passes a certain threshold. The time the test lasts is called the wall's fire rating. Obviously, the higher the fire rating the longer the wall contains the fire. The fire rating is rounded down to the nearest hour or quarter hour, so you get fire ratings like half an hour or 45 min, not 37 or 52 min.

Depending on exactly what sheathings are put on wood frame walls, they routinely get fire ratings of $\frac{1}{2}$ to 1 h in the tests. And by the way, they are almost always structurally useless at the end. They cannot support a significant weight anymore.

A few ICF companies forms have paid to have their walls tested, too. Even the walls with the thinnest concrete inside and no finishes over them routinely last 2 h. When finishes are added and the concrete is thicker this may increase to 4 h (see Fig. 1-11). And the concrete wall is almost always structurally intact afterward (see Fig. 1-10).

Resistance to Other Disasters

Other disasters are rarer than hurricanes. So we do not have similar studies of building survival for these. However, the information that's available suggests that ICF buildings hold up well in them, too.

Earthquakes

José, a longtime ICF contractor in California, tells about his experience with an ICF house in an earthquake:

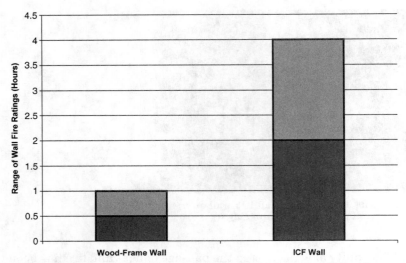

Figure 1-11 The range of fire ratings for typical wood frame and ICF walls.

I was at the top of a three-story house I was building about 5 years ago. I was putting the top plates on. A 4.5-rated earthquake hit. What I felt was one big thump. It just felt like a huge thump and that was it. It felt like a truck had run into the side of the building, lower down. But I looked at the stick-framed house next door, and I could see it moving in a wavy motion. It was wiggling and cracking, and you could hear it popping.

After the quake was over, I did a very thorough examination of the structure of my house, and I didn't find any significant damage, in fact no damage at all. I checked out the house next door and it had nonstructural damage, but still significant in repair costs. It had some wood flexing damage to sheetrock, doors, and windows. The usual things—doors and windows flexed out of alignment, cracks in the sheetrock from the corners of the jambs.

That same ICF house I built has been in other earthquakes since then and it still hasn't had any significant damage.

Several ICF homes have now gone through earthquakes, and in some cases the quake measured 6.0 or higher on the Richter scale. In all cases their owners reported that there was no significant damage. The key here is that the rebar used has to be adequate for the local seismic conditions.

Properly reinforced concrete has historically survived earthquakes well. It is true that very old buildings constructed of unreinforced concrete (no rebar inside) have suffered significant damage in quakes. The shaking cracks them, sometimes, into pieces. But modern concrete buildings in high-seismic areas follow new building codes that require rebar, and more of it.

Floods

The forces put on a building by the surging water of floods are similar to the forces of high winds. So it is logical that ICF walls should stand up well to floods, as they stand up well to hurricanes. A few owners of ICF buildings report that they have gone through floods, and all say that the buildings stood up without structural damage. In addition, the ICF walls did not rot as wood might.

Explosion

Just in case there was any doubt about the ability of ICFs to withstand severe forces, the military is trying to blow them up. Every so often the U.S. Department of Defense conducts its Force Protection Equipment Demonstration (FPED) at Quantico Marine Corps Base in northern Virginia. This series of blast tests shows how various products perform in an explosion. The latest FPED, in May of 2003, included six "boxes" of ICFs. These were cube-shaped structures that measured 8 ft × 8 ft × 8 ft, or about the size of a small room. According to Joe Lyman, the Executive Director of the Insulating Concrete Form Association, who observed the tests:

> We blasted the boxes (8' × 8' × 8', concrete at 4000 PSI, 6-inch slump, $^3/_8$" aggregate pump mix, horizontal and vertical rebar at 16" on center) from 10 feet away and 6 feet away on the final day with a 50 lbs charge of TNT. The boxes experienced very limited cracking (less than 2 millimeters across) with no structural damage at all. In fact, the boxes didn't even creak when they were craned onto the flatbed to take them off site.
>
> All three days, we experienced the impact resistance the foam provides. The box blasted 6 ft away, experienced about 275 lb of pressure per square inch, and our blast designer said we should have had a hole in the face of the box at least 1 ft in diameter with major deflection. However, it survived with limited damage. It appears that on all six boxes the foam compressed against the face of the concrete and absorbed the brunt of the blast (see Fig. 1-12).

Sound reduction

Especially as development gets more dense and more buildings are close to streets, rail lines, and airports, the amount of noise that enters from outside is a growing concern. ICF walls have excellent sound attenuation.

Today, scientists measure the amount of sound that gets through a wall with the **sound transmission coefficient** (STC). The idea is pretty simple. The testers make sound on one side of a test wall and measure how loud it is on the other side. The greater the reduction in the sound energy, the higher the STC and the better the wall is at stopping sound.

In the 1960s the National Association of Homebuilders Research Foundation tested the STCs of some typical frame walls. The results are still valid because the materials they used—2 × 4s with fiberglass insulation and gypsum wallboard on each side—are virtually the same today. They found that these walls had an STC of about 36 (see Fig. 1-13).

Some of the ICF suppliers have paid to have their walls tested, too. Depending on the brand of ICF tested, the thickness of the concrete, and so on, these have shown an STC of 46 to 50. When you add finishes, that increases.

Figure 1-12 ICF "box" after surviving explosion of a 50-pound charge of TNT nearby. (*Courtesy of ICFA.*)

An STC that is higher by 10 points may not sound like a lot, but when you understand the way STC works you see that it is. Without getting into the details, every time you increase the STC by 10 you cut the amount of sound energy that goes through the wall by a factor of more than three. So less than one-third as much sound gets through a wall with an STC of 46 as gets through a wall with an STC of 36.

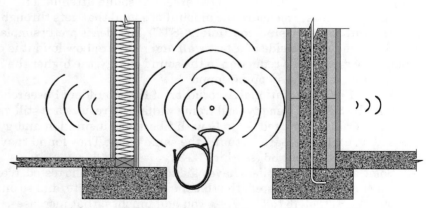

Figure 1-13 A frame wall (left) stops less sound than an ICF wall (right).

Here's a more intuitive way to understand the difference. Tables published by material suppliers show that loud speech on the opposite side of a wall comes through as "audible but generally not intelligible" if the wall is STC-35. But if the wall is STC-48 the listener "must strain to hear it," and if it is STC-50 it is "inaudible."

The owners of ICF houses realize the difference quickly. Many of them tell the same two basic stories. According to two particular owners:

> When we toured the house I stopped at the front window. I could see the cars passing by on the main road, but I couldn't hear them.

> One morning I was out in the yard and I started visiting with my neighbor. He said, "Man, did your kids start crying when that storm hit, too?" At first I didn't understand, but we talked some more and I figured it out. There had been a thunderstorm the night before and it had woken them all up. But we didn't even realize there had been a storm at all.

One word of caution is that there are other avenues for sound to get into a building than just the walls. It also comes in through the roof, windows, and doors. These things may be about the same in a frame or ICF building. If so, they will reduce the sound attenuation difference between the ICF and frame walls some.

Durability

Structural concrete has some durability advantages over wood frame. It does not warp or rot. In fact, it is pretty much totally unaffected by water and humidity. It does not settle or shift. It cannot be eaten by insects or vermin. Yet all of these things can and do happen to wood structures from time to time.

The one warning here is that it is important to take proper precautions against insects. Manufacturers and code officials have been careful to take steps to prevent insect problems in the foam of ICFs. You need to be sure you don't short-circuit them.

Termites have been known to tunnel into foam that is in contact with the soil. There appears to be no documented case of termites getting into the foam of an ICF wall below ground and tunneling its way up to wood, where it could have a feast and do real damage. However, as a precaution, the pest control industry, model building codes, and the foam and ICF industries decided that there would be certain restrictions on the use of foam below-grade in high-risk areas.

The restrictions are now written in the International Residential Code and the Standard Building Code (which some people call the Southern Building Code). They show what the very heavy risk areas are on a map. They are mostly the coastal states of the Southeast, plus most of California and all of Hawaii. These are where termites are most active (see Fig. 1-14).

The rules allow buildings to have ICFs extending below grade in the high-risk areas. However, if they do, some other step to deter termites must be taken. At this time, there are three steps that are accepted termite deterrents. The first

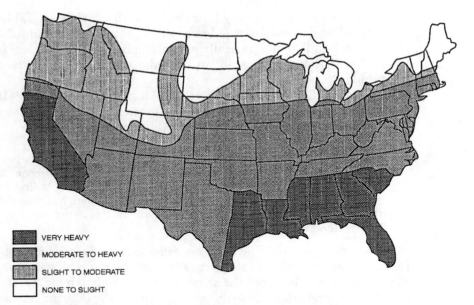

VERY HEAVY

MODERATE TO HEAVY

SLIGHT TO MODERATE

NONE TO SLIGHT

Figure 1-14   Termite risk areas of the United States. (*Courtesy of International Code Council.*)

one is to use ICFs made of borated foam. Borates are naturally occurring minerals that kill termites but are nearly harmless to animals and people. Borated foam is just foam that has borate powder mixed in. Some ICF manufacturers supply their forms with borated foam.

The second accepted termite deterrent method is to cover the foam with a particular type of waterproofing membrane. Manufacturers' tests show that termites cannot readily penetrate this membrane. So putting it over the foam stops them before they get to the foam.

The third method is to cover the foam with a special stainless steel mesh. This has also been proven to be a layer that termites cannot easily get through.

Over time there will probably be more and more accepted measures. The code bodies have a committee that reviews new measures that come along and accepts the ones that prove they can do the job. Many companies are now developing and testing their own products to stop termites from tunneling in foam. So keep your eyes open.

There is one other way to "termite-guard" the building. If there is no structural wood in the building, there is relatively little damage termites can do to it. So the code requirement to install an extra termite deterrent does not even apply to a building with no structural wood. Eliminating all the structural wood is a real possibility—a growing number of ICF houses have interior walls, floors, and roofs constructed out of light-gauge steel framing or even reinforced concrete. They all qualify.

Remember, too, that these requirements only apply if the building has foam extending below grade. Many buildings in the high-risk areas are built on a slab. There is no basement or stem wall. The bottom of the walls starts several inches above grade. If the gap between the walls and the ground is sufficient, there is no requirement for extra steps in these cases.

And rest assured that all of this is a precaution only—termites have not been observed to do significant damage to ICF buildings. And they will never be able to damage the concrete that forms the wall structure.

Indoor Air Quality

The trend in construction is to make tighter and tighter buildings to keep out unwanted pollutants and moisture, and then bring in controlled amounts of fresh air through the ventilation system. This approach is a bit easier with ICFs than many other wall systems because they are quite air-tight without the contractor having to take any special measures.

In tests, ICF houses have shown air infiltration rates around a third less than the rates of frame houses. The difference would probably be greater, but the infiltration through other parts of the shell like the roof and openings can still be the same as in a frame house. The ICFs virtually "seal" the walls, so how tight a house you get depends on how you handle the other parts.

The manufacturers also like to point out that ICF walls do not provide a food source for mold. To grow, mold requires a certain level of moisture, a certain range of temperature, and a food source. The food must be some organic material. And wood, for one, is organic. So with the right moisture and temperature conditions mold can and has grown on wood. This is particularly a problem when water gets into a frame wall. But there's nothing organic in an ICF wall—not the foam, not the concrete, not the rebar. If any mold were to grow on it, that would have to be where some organic material had settled on the surface. And that is likely to be in very small quantities.

But a word of caution: in *any* building, assuring indoor air quality takes some attention to detail. If the building is constructed to be very tight, the HVAC system should be designed to introduce proper amounts of fresh air and keep humidity levels from getting too high. Exterior details need to keep water from getting inside.

These are all important measures that go hand-in-hand with constructing ever-tighter building shells. The advantage of ICFs is that they help you get to the tighter shell a bit more easily than other wall systems. And they stand up better to any moisture that might accidentally be getting in.

Comfort

The same things that make an ICF building energy efficient also keeps the interior more comfortable. There are fewer "cold spots" and "hot spots." The temperature is more stable. And there are fewer drafts. These things are somewhat subjective. But occupants notice them once they have been in an ICF building for a while.

Cold and hot spots along the wall are almost nonexistent in an ICF building. In a frame building, the locations in the wall that contain wood do not have insulation. At places like corners and at openings there can be several studs next to each other. These create large uninsulated areas where the wall will be colder in winter and warmer in summer. ICF walls have a nearly unbroken layer of insulation all around.

Besides reducing overall energy consumption, the thermal mass of the walls helps keep the indoor temperature more stable. It takes considerable time for the concrete in the walls to heat up or cool down. This helps to even out the highs and lows of temperature that result from outdoor temperature changes or from the on-and-off operation of the heating and air conditioning.

The lower air infiltration through the walls, of course, means fewer drafts. In sum, the indoor environment is much more stable—both from time to time and from spot to spot.

Sustainability

In the last five years, consumer interest has mushroomed in the impact that their products have on the natural environment. Terms like **green building** and **sustainable architecture** are now widely used. People's definitions of these terms vary, but generally speaking the goal behind them is to disrupt the environment as little as possible.

A complete accounting of all impacts from any human activity is impossible, but some research has been done on the amount of energy used by ICF houses, and the emissions of carbon dioxide (a so-called greenhouse gas). These two are related because using energy almost always produces carbon dioxide.

According to a study from the Cement Association of Canada, the total consumption of energy and the emission of carbon dioxide associated with an ICF house is typically much lower than with a wood frame house. The initial levels of energy and carbon dioxide from making the building materials are higher because cement, one of the major components of the concrete, uses a lot of energy in its manufacturing process. But after that the ICF house uses less energy for heating and cooling than the frame house does. Within 2 years the ICF house has saved as much energy as the extra energy required to build it, and the savings after that just continue to accumulate. The same goes for the total carbon dioxide emissions. So over its entire life cycle, the ICF house consumes much less energy and produces much less carbon dioxide.

The foam in ICFs is a petroleum product, and most people consider petroleum a nonrenewable resource. On the other hand, the foam stays in the building and serves a permanent purpose—insulating the building. It is not simply burned, the way oil used for fuel is burned. In fact, over its life the foam saves an amount of fuel that is much larger than the amount of oil it contains.

Lower Insurance Rates

Most insurers have a special rating category for buildings constructed with materials that are more resistant to fire and structural damage than conventional construction is. A typical story comes from Chris, an ICF homeowner in Ontario:

> I realized an approximate reduction of 25%–30% in my home policy. But I would like to explain that it was not due to the I.C.F. per say, rather it is what's behind the "foam"-concrete. My agent simply looked in his guide and entered the applicable rate. A "stick built" home will cost more for structural insurance than the concrete home. In the concrete home, the inevitable payout from insurers due to fire, flood, storm damage, etc. has been shown to be far less.

You may have to call the agent's attention to the concrete core to get the favored rating. And you may have to shop a little—not all companies offer this reduction, and some may offer different savings.

Better Mortgage Terms

Many home mortgage lenders offer **energy-efficient mortgages**. These are sometimes also called **stretch mortgages**. They allow the owner to borrow more than lending rules would normally allow.

These mortgages are for houses that meet higher-than-normal energy efficiency standards. The lender reasons that if the house uses less energy, the owner will have a lower monthly energy bill. That means that the owner will have some extra money to devote to paying off the mortgage. The lender might normally not want the monthly mortgage payment to be more than, say, 28 percent of the borrower's monthly income. But with an energy-efficient home the lender approves a mortgage payment as high as, say, 31 percent of income. This means that the owner can borrow more money.

This may allow some people to buy a home they otherwise could not afford. For example, it might permit the buyer to qualify for a house with a couple of hundred extra square feet. Or, there might be someone who can qualify for a frame house, but who can't quite qualify for the same house built with ICFs because it costs a little more. With the stretch mortgage, the energy savings of the ICFs might allow that person to borrow enough to cover the extra cost.

There are different ways the borrower can prove that the house is in fact energy efficient. You have to follow the rules of the particular lender. But houses with ICF walls almost always qualify, so long as the rest of the construction is reasonable quality.

The greater challenge with a stretch mortgage is finding a loan officer who knows how to write them. As we discuss later, they are rare enough that many loan officers are not even aware that their company offers them. So the borrower may have to be persistent to get one.

2

Benefits to the Contractor

Overview

It's great that ICFs create a better building for the owner, but there's a lot more to weighing the benefits and deciding whether they make business sense for the contractor.

Tom, a technical advisor for one of the ICF companies, gives a rapid-fire delivery of potential advantages to the contractor:

> Some of the advantages are expanded market, more business, and more money. Then there's speed of construction—time equals money, saving time allows for more projects, there are fewer sub trades, there is less risk of site injuries due to heavy lifting, and it's easy to do accurate material and labor estimates.

And this is not the end of the list.

You have to think about how important each of these things is in your case, and weigh them all against the time and cost of learning something new. This chapter walks through major things to be considered when deciding whether ICFs are for you. It is brief, almost a checklist. The later chapters go into more detail on many of these same subjects. This one gives you the heads-up of what to be thinking about as you read. By the end of the book you should have a good base for deciding what you want to do.

Selling Price

Randy, a manager with a large Texas homebuilding company, tells about the company's experience with a development of ICF houses it constructed and marketed a few years ago:

> We sold the houses at a premium of $3.50 per square foot. They all sold quickly. We haven't done nearly enough work to know how much you can sell ICF houses for. We had to set a number, and we did it with information from focus groups and appraisers. Maybe we could have charged $4, or maybe we should have charged $3.

Other homebuilders have similar stories. Their ICF houses can command a premium. In fact, since most buyers of ICF houses come to the market looking for ICF construction, they are usually aware that they'll be paying a little more.

But don't assume that you can automatically get a hefty premium. Randy's company did a lot of promotion of the benefits of ICFs to educate the consumers about what they were getting for their money. There are more and more promotional campaigns to educate the public that will help you with your sale. But in some areas awareness is not as high as in others. Especially in these areas, you will either want to rely on the buyers who have researched ICFs themselves and have decided they want them, or you should be prepared to explain and promote the benefits of ICFs to the buyers.

In commercial construction the situation is a bit different. There ICFs may be less expensive than competing wall systems, depending on the type of building and its requirements. So on a commercial building you can operate pretty much according to standard bidding procedures.

Cost

Most people agree that the cost of constructing a building with ICF exterior walls is a little higher than the cost of constructing it with standard frame walls. But not everyone. Lon, a longtime homebuilder in Oregon, says:

> I've come up with an efficient operation over the years. Right now I can build an ICF house for about the same price as frame, maybe even a little less. It's because the ICFs give me some savings on other parts of the house. The ICF walls themselves are definitely some more, but they save me money other places.

In commercial construction, whether ICFs are more or less depends heavily on what you're comparing them to. In a lot of projects the other wall system that would have been used if ICFs weren't around would actually be more expensive, so the ICFs offer the contractor and the owner a less-expensive option. If all the owner needed was basic light storage good for a few years, a light metal building might do the trick, and the cost of that would be tough for anything—including ICFs—to beat. But as soon as the job starts requiring some higher structural strengths, energy efficiency, sound deadening, and the like, ICFs can be very cost-effective. These attributes come standard with ICF walls, but outfitting a conventional wood or steel or concrete wall to have all these features can require a lot of added time and materials that jack up the cost of the other wall systems.

Estimated costs

In homebuilding, typical wall construction costs for an ICF subcontractor currently run about as follows:

Item	Cost/gsf of wall area (US$)
Forms	3.00–3.50
Concrete	0.75–1.50
Rebar	0.20–0.40
Buck material	0.05–0.25
Misc. materials	0.05–0.10
Pump rental	0.00–0.40
Labor	1.00–2.00
Total	5.05–8.15

The subcontractor of course takes a markup to pay for his acquisition of materials, management of the project, and so on. Typical rates charged by ICF subcontractors to general contractors currently run about $6.50 to $8.50 per square foot of gross wall area.

The "gross wall area" that ICF contractors talk about is the area of the exterior walls of the building, without taking out anything for openings. They work this way because it gives the most accurate quotes.

Some conventional contractors estimate their costs per square foot of *floor area*. But this is very unreliable for ICF work. For example, if you had a house with 8-ft walls and you changed the design to 9-ft walls, the cost of the ICF construction would go up by around one-eighth; about 12 percent. But, since the floor area stays the same, anyone who quotes by floor area would not change his price.

The per-square-foot-of-floor-area rule makes some sense for light frame construction. The height of the walls is pretty constant, it takes about the same amount of work to nail up a 9-ft stud as it takes to nail up an 8-ft stud, and in most cases the framing subcontractor doesn't pay for the materials. The general contractor buys the lumber, so if the amount of lumber changes, it figures on the GC's costs, not on the framer's costs.

So be careful when comparing the costs of ICF construction to the costs of other walls. It is best to take total costs for the whole job so you will be certain of comparing apples to apples.

The wide range of the costs that are quoted results from a hundred different little factors, like with any other type of construction. In fact, for really crazy types of buildings you could easily get higher than the figures quoted above, or lower.

The price of the formwork is pretty constant. You might buy bargain forms from somewhere, but then other costs from things like waste and rework and labor time are likely to be higher. In a project with a lot of special wall features like non-90 corners and brick ledges, the form cost can be higher because of the need for a lot of specialty form units.

The concrete costs depend a little on your local ready-mix prices and a lot on the thickness of the walls. The low-end price corresponds to forms with a 4-in cavity, and the high-end is for 8-in forms. The 4-in cavities are used here and there for above-grade walls. The 8-in forms are used mostly in basements and above-grade walls in some commercial buildings. Most above-grade residential walls are now 6 in, and some basements are. That will of course lead to a concrete cost that is in between. Now and then you find thicker walls, mostly in commercial construction. The concrete for those will be even more.

Rebar cost can vary all over the map because of big differences in structural requirements. In most areas foundations require more bar, so the steel cost is higher for a basement. In locations with low wind and earthquake risk, the plans might call for one vertical bar every 4 ft, while in high seismic areas they might call for one every foot, which jacks up the cost of rebar fast. However, as you can see, it is not a large component of cost.

Bucks are the frames built around window and door openings. They are constructed of wood or special plastic extrusions and left in place for attaching the window/door, trim, and so on. Total cost of the material depends mostly on the number and size of openings, so it will be very low in things like foundations and certain uninhabited buildings, but high above-grade in residences, offices, and so on.

Miscellaneous items such as screws, reinforcing tapes, adhesives, and so on add a little cost, but most contractors don't even keep track of it.

A few contractors pour their concrete from the ready-mix chute in below-grade jobs. So the cost of a concrete pump is zero. But many now use a pump even below grade, and all do so above grade. The cost runs about 20 to 40 cents per gross square foot (gsf). The difference depends, mostly, on how efficiently the pump is used. If there is nothing available but a large boom pump for a small foundation, cost will be at the high end. If it's possible to use a small pump for a small job, or bring out the large pump for a large pour, cost will be near the low end.

Labor varies widely because complexity varies widely. A reasonably experienced crew can typically do all the work for an ordinary above-grade story of a house at the rate of about 20 square feet of gross wall area per each worker-hour. That includes the entire ICF job, from marking the locations of the walls on the footing or slab, through building bucks and stacking forms and setting up bracing/scaffolding and aligning the walls and putting bucks and rebar and any inserts in the walls and placing the concrete, to cleaning up the job site after the pour. If the job is extremely simple, with four 90° corners and few or no openings, productivity could go up to 40 gsf/worker-hour. For a story with many corners, irregular corners, lots of openings, geometric openings, and curved walls, the rate could easily drop to 10 gsf/worker-hour.

The other big labor variables are crew experience and site conditions. On its very first job may require 50 percent more labor to do a project, and a crew that has worked together on 20 ICF projects may take less than the figures quoted here. Bad conditions drive down productivity and drive up cost; consistent sunny days with temperatures of 70°F do the opposite. Things like site access also play a role.

Commercial construction costs can differ from residential ones just because the walls differ. In, say, a small office building the costs will be similar to residential because the complexity of the walls and the amounts of concrete and rebar will be similar. But walls for a large storage building with long runs and very few openings are so efficient to build that they might even be less expensive than typical residential foundation walls. Costs will start creeping up when you get into very high walls that require adjustable scaffolding and thicker concrete with higher reinforcement levels.

Total building cost

The impact on total building cost also varies, but there are more and more good data on it. In residential construction, the popular rule of thumb is that a house with ICF walls usually has a total cost that is about 1 to 5 percent more than it would have been if the exterior walls had been built with frame. So a house built with frame for $200,000 would cost $202,000 to $210,000 with ICFs.

But the difference appears to be narrowing, and some contractors claim it is at or near zero for them. The cost of the walls is coming down as the manufacturers and contractors become more efficient, but the even bigger factor is the potential for savings in other parts of the building and other parts of the construction process.

If the heating and cooling equipment is sized correctly, it can typically be much smaller in the ICF building. Reductions of $800 to $2000 for a single-family house are pretty common. Using ICFs often makes it possible to have one crew responsible for the foundation and the above-grade walls and the framing of the interior and the roof. This can eliminate a lot of time spent waiting for a foundation crew to come in and out, with a resulting reduction in the builder's finance costs. New savings are appearing as crews and municipalities get more used to the product and see that it doesn't always require all the steps and measures that are taken with frame construction.

With commercial construction similar savings in other parts of the building may be possible as well. For example, in a building that requires precise environmental control, the ICF walls can save huge amounts on the HVAC system. If the ICF wall system is not already less expensive than the competitive wall systems, this type of savings might make the total building less expensive and swing the cost issue in ICFs' favor.

Local Market Potential

ICFs are now built into about 4 percent of all single-family homes constructed in the United States and Canada, and sales are growing a breathtaking 30 percent per year. They are a smaller share of low-rise commercial construction, but they appear to be growing even faster there.

Still, you have to ask yourself how big your local market is. To be successful with ICFs, you need to have a reasonably large pool of local customers who could be interested in the benefits that ICFs offer, and are *willing* to pay for them.

In some ways this is not asking a lot. The record shows that there are plenty of buyers who want ICFs as soon as they learn about them. But you don't want to jump into this market without some idea of where the business is going to come from.

Look around

First and foremost, it pays to see what's going on with ICFs in your area. Who is using them? What are they building? What do the owners say about the projects? If ICFs are taking off, it's pretty clear your area has potential. You may be a little late to the market, but you know the market is there. If they haven't taken off yet, things may be riskier, but you could also have the market to yourself.

A good way to find out what is going on locally is to contact the major ICF suppliers and ask for the names of their local distributors. You can get contact information on the suppliers at the web site of the Insulating Concrete Form Association (*www.forms.org*). The suppliers are in the Directory under *Primary Members*. Once you get hold of their distributors, these people will gladly tell you about their nearest buildings, since you are a prospective customer. They might even give you the name and number of the buyers or the contractors themselves so that you can ask for their views. And the contractors, unless they consider you a competitor, are usually happy to tell you about other ICF jobs they know of.

There are also plenty of directories that give you the names of local ICF contractors directly. Many contractors are members of the ICFA, so they are in the ICFA web site directory. The independent web site *www.icfweb.com* also has a directory that lets you search for contractors by state. The people who run ICFweb also publish a print version of their directory that they sell. And Concrete Homes magazine has state-by-state listings at the back of every issue.

But what if there is little or no ICF construction in your area? That doesn't prove that ICFs won't sell. It might simply mean that no one has tried enough to sell them before. Some areas have been faster to adopt than others, but every week ICF construction is getting established by someone in a new town somewhere.

If there is limited ICF building near you, you need to figure out whether there are people who are *likely* to buy if you offer the product. The local distributors are good allies in this case, too. Your town might be one they have been trying to open up, and they may know of some interested buyers or where to look to find them. You might also check the project listing services, like the one on *www.icfweb.com*. Buyers list buildings they want bids on with ICF construction. If you find one or more in your area, you have a clue that there is a demand for ICFs that no one is filling. Depending on the rules of the service, you might even be able to register and get in touch with the buyers to find out more about why they want ICFs.

One final way you can research how well ICFs will sell is by figuring out who is likely to buy and simply talk to a few of these people. At a party or at church or in any gathering, explain the product to a few people informally and see how they react. But don't talk to just anyone. The record shows that there are certain groups that are much more interested in ICFs. To be specific, the current buyers of ICFs are concentrated in *high-end homes, certain low-end projects*, and *low-rise commercial buildings*.

The high-end housing market

According to Ray, an ICF supplier in New Mexico:

> The typical buyers earn over $70,000 per year and live in a selective community.

High-end homes sell to the wealthy, and these are people who want better things and have the money to afford them. You often hear that homebuyers will not pay more for anything that exceeds local code, or anything that they cannot see. That may be true, but only to a point. As they get more money, people may first spend their money on Italian tile and an elegant master bath. But after a point they have been willing, time and again, to devote a few thousand dollars more to a superior structure when they understand it and appreciate what is better about it. So if you are the type of contractor who can build upscale homes and likes to talk about his products, this is a market that may make immediate sense for you.

The mid-range housing market

Nine times out of ten, buyers in the middle of the housing market can't get past the issue of the first cost of their home. Even if you could show that once they're in the house their insurance and utility bills will be lower, they cannot bring themselves to pay a little more for a house up-front.

But bear in mind that the cost of everything in ICF construction—the forms, the amount of labor, the tools—is coming down, slowly but surely. At the same time, building requirements like energy codes and wind codes are getting more stringent, and that is pushing up the cost of frame construction. As the prices of ICF houses edge down and frame prices edge up, people in the middle and lower-end housing markets will start buying more ICF houses.

Some ICF sellers already claim to be witnessing significant sales in the middle market. According to Will, a distributor in Georgia:

> We're seeing regular people. It's not all big houses. It tends to be people who have enough equity in their existing home or enough savings that they can get into an ICF house. It tends to be people who are a little more sound financially to start with. But they aren't all rich.

The low-end housing market

Interestingly, there are also some good opportunities here and there in affordable housing. A great example comes from the city of Lubbock, Texas. The city sponsors the construction of several dozen houses each year that are partly subsidized by government programs. According to Brad Reed, Senior Building Inspector and Supervisor for Community Development:

> As an experiment, we contracted with local builders to construct some ICF houses and we measured reductions in utility costs. The savings came in at over $30 per month for a typical house. I used to offer the owners the option of either ICFs or wood frame. But [with the subsidized loan] the ICFs only increase the monthly

mortgage payment about $4, so with the $30 utility savings everyone started to choose ICFs. Now we don't even offer wood. All our affordable housing construction is ICFs.

Other low-cost housing programs have made a similar changeover. In all cases, a major factor is that the total monthly payments that the new owners have to make are lower with ICFs because their savings in energy consumption and insurance offset the higher mortgage cost. And with the up-front cost often subsidized, that cost is not so important to the buyer.

These programs typically hire local contractors to do the work. If programs like these in your area currently specify ICF construction, or are willing to consider it, this might be an opportunity for you.

Commercial markets

ICFs are new to commercial construction, but are growing rapidly there. They are popular in many different types of commercial buildings, but not a *random* set of buildings. Almost all commercial ICF buildings hold people or some other contents that require a controlled environment: offices, hotels, retail, churches, wineries, movie theaters, refrigerated storage, and so on. Clearly, ICFs' energy efficiency and sound reduction are the attraction.

ICFs excel in low-rise commercial buildings, up to about 40 feet or five stories. The lightness and flexibility of the forms are very efficient in these heights. They have also been used as the curtain (nonstructural) walls of heavy steel-frame high-rise buildings up to 11 stories. However, their advantages are usually greatest at the lower elevations.

In fact, ICFs may often be less expensive than alternative wall systems in commercial construction. Commercial buildings frequently have heavier loads, so the walls have to be beefier. For frame construction, that leads to a lot of extra wood and steel connectors, which drive up the cost. ICFs can often meet the requirements with nothing more than a little additional rebar. Heavy steel construction and other forms of concrete construction may end up being more expensive once you add other things the job requires, like insulation and furring.

So if you work well in a commercial environment and you are open to building a variety of structures, ICFs may open up a good chunk of that work for you.

Construction Advantages

Using ICFs can give you, the contractor, several valuable logistical advantages.

Reduced cycle time

According to Buddy, an ICF contractor in North Carolina:

It used to be that we did the footings, then we'd wait for a mason, and they would eventually come in and start work, but then they'd have to take off for a while, and so on. You could easily lose a month on the foundation. But now with ICFs our same 3 guys who did the footings can start the basement immediately afterward. They can stack an average basement in 3 days and pour it the next day and start doing the walls on top of that the day after. You can easily pick up a month on a job this way. Plus the framer never had to leave and find other work.

For the general contractor, a reduced cycle time means paying less interest for the construction loan, and that lowers costs and increases profits on the project. For *both* the general contractor and the subcontractor, the lower cycle time makes it possible to complete more projects in a year, and that increases total revenue and profit.

A bigger share of each project

ICFs can give you more of the job. If you are already a framing contractor, in an ICF building you can bid on the foundation (basement or stem walls), the above-grade exterior walls, and all the interior and roof framing. If you are a foundation contractor, you can bid on all the exterior walls right up to the roofline, not just the ones below grade.

Less coordination

If you are a general contractor, you can sometimes eliminate one crew that you have to deal with and coordinate. You can simply hire one framing crew to do the entire shell from the bottom of the foundation to the top of the roof. In that way you eliminate the masonry or forms crew that would otherwise have done the foundation. In a lot of areas today this is very attractive because masonry crews are becoming hard to get, hard to coordinate, and expensive.

In both residential and commercial construction, you will reduce your coordination with the insulation contractor, and you may eliminate the need for him altogether. The ICF walls will already be insulated. If you happen to do an ICF roof, all the exterior insulation is taken care of, and there is no need to bring out the standard insulation crew at all.

Job site workload and safety

Most of the work at the job site is with foam, which is a light, soft material. Crews need to do less heavy lifting so that they are fresher at the end of the day. There will be fewer injuries from muscle strain or from building materials accidentally striking a worker.

Less waste and cleaner job site

According to Jeff, an ICF contractor in Maine:

> On a big basement, we get our ICF scraps in one large trash bag. One guy walks it off the site and we're good to go. Everything else we have to remove is equipment.

It's easy to use cutoffs and odd pieces of the forms by gluing them back into the wall. Concrete, the other high-volume material, is typically ordered when the formwork is up in an amount that is just enough to fill it. On a typical pour for a house, maybe a yard of concrete will be left over and the crew can usually find a use for that in a garage apron, a sidewalk, or a pad.

Commercial

In commercial construction also ICFs often have advantages in flexibility of design and speed of construction. This depends a lot on the project, because each job has different requirements.

The speed issue can be even more important in commercial construction than in residential. If a house is ready a month early the owners just get to move out of their old place sooner. If a store is ready a month early, it opens for business and generates maybe a few hundred thousand dollars in sales that the owners wouldn't have gotten otherwise.

Cost of Entry

To become an ICF contractor you will probably need the following:

- Training
- A few new tools
- Access to a bracing/scaffolding system

That's about it.

Training is critical for the lead person on the installation crew. That is the person who will be telling others what to do and checking their work. The rest of the crew is typically much lower-cost workers. And in most parts of the country they turn over pretty frequently, so it's hard to keep them fully trained. If the lead is good that won't matter much.

Training is also excellent for crew supervisors and managers of larger companies. Even if they won't be setting forms, they need to know what the crews are up to so they can supply them with the right tools and evaluate their work.

This book covers training options in detail later. The most common way to get training is through a formal course from one of the ICF manufacturers. Fees range from $200 to $500 per person. The courses run two to three days and include classroom training, hands-on exercises, and sometimes site visits. You will never get a perfect feel for the products until you handle them in the field, but these courses give you a big leg up. They also give you the opportunity to talk with other experienced and novice ICF contractors. This can be an education in itself.

For the most part ICF construction uses common tools that every framer already owns. But sooner or later you will probably end up buying a **rebar cutter/bender**, a **hot knife**, and a **foam gun**. These are extremely handy. They save a lot of time and improve the quality of your work in three important tasks—cutting and

Figure 2-1 ICF bracing/scaffolding system in place. (*Courtesy of Insulating Concrete Form Association.*)

bending your steel rebar, making neat cuts in foam, and applying adhesive to glue forms together. They are discussed in more detail in Chap. 6. Depending on how fancy you get, you can buy all these things for around $500.

Critical to working fast and doing quality work is having a good **bracing and scaffolding system** (see Fig. 2-1). This has two functions. It gives the workers something to walk on for the high-up work and it keeps the walls plumb and straight until the concrete cures. Since the number one quality control issue with ICFs is producing straight and plumb walls, you really don't want to fool with a poor bracing system or makeshift measures. Concrete is not a forgiving material once it is in place, so you need to do things right the first time. There are two kinds of contractors who do not spend the money for a good bracing system and spend the time to use it right. They are the dumb contractors and the dumber contractors.

When you're getting started it's a good idea to rent the bracing/scaffolding equipment. Most local ICF distributors can supply it to you. This saves you from having to make a big investment up-front before you have things rolling. It also allows you to try out some systems so that you can decide which type you like the best. In fact, some crews find it most economical to rent the scaffolding/bracing equipment for every job.

But if you start doing a steady volume of ICF work you will probably decide to buy your own equipment. It usually works out to be cheaper, you always have the equipment around when you need it, and you can get used to one system. A typical bracing/scaffolding system good for residential construction costs $50 to $150 per set. The higher price gets you more durability and adjustability. One set is erected about every 6 ft along the inside of the ICF wall during construction. Depending on the size of your business you might need between 20 and 40 sets. There are also ways to construct your own system by using standard 2 × 4s and attaching brackets and turnbuckles to them. They cost about $30 to $40 per set in materials. They can be used multiple times, but tend to be a little heavier, slower, bulkier, and less precise than a factory-built system.

For large commercial projects you may need a heavier system that is designed to support walls in the 10-foot-plus height range. You will also need more sets. All of this will cost you more. But the payback can be fast because of the time savings a good system provides, not to mention the improved quality of the work.

Learning Curve

While it may take 2 years to become a good mason, it takes three to five jobs for an ICF crew to become proficient. After that they will continue to improve, but much more gradually. It's typical to build with about one-half to two-thirds as many labor hours on the fifth job, compared to the first or second.

Even though it's short, there is a definite learning curve and it is foolish to think otherwise. As Jim, an ICF contractor in Connecticut puts it:

> There are no reasons to accept improper installations of any product. Manufacturers provide training or site assistance with their product, yet some installers (homeowners or contractors) fall through the cracks for simple reasons. They don't respect the products involved and are convinced that this is a simple process. Just stack and glue, anybody can do it. Yeah, right! You get maybe 45 minutes working time with concrete; you don't get to fix it easily the next day.

Before your crew gets experienced, you need to consider how they will get down their learning curve. The most safe and proven way of learning the basics of ICF construction is to go through a formal training course and then have an experienced contractor help you with the first one or two jobs. Many times your local ICF distributor will send an experienced person to your job site. Or you can find a nearby contractor who has built a few ICF buildings and get the lead man out to help. If it costs, pay it. You'll save time and money on the job to cover the cost. And you'll perform faster and better on every job after that.

Design Flexibility

With ICFs you can construct practically any design that you could with frame or other materials. Some things are even easier to do. Because the foam is light, easy to cut, and bendable, you can often create curved walls and irregular openings like arched windows and doors faster and easier. The strength of reinforced concrete makes it possible to form long-span headers for openings. And ICF floor systems can make long clearspans (see Fig. 2-2).

A few design features are harder with ICFs. Mostly these are spans and projections—overhangs where some of the walls of an upper story don't line up with the walls below, but just float in space. Most houses don't have them, and if they do they can usually be designed out. But if you're in an area where

Figure 2-2 ICF foundation (top-left), high-rise (top-right), and house (bottom) with curved walls and irregular corners. (*Courtesy of IntegraSpec, Arxx Building Products, and Quad-Lock Building Systems Ltd. respectively.*)

cantilevered upper floors are common and the buyer will insist on them, be aware that you may have some added work and cost to build with ICFs.

ICFs also lend themselves well to change orders—if they're reasonably early in the construction process. Cutting up walls, moving openings, and shifting footprints during construction are very easy when you are working with foam. It is mostly a matter of cut and paste. But this flexibility goes down fast when the concrete is poured. Changes can be made—and sometimes are—by getting a concrete cutting crew out after the walls are hard. But changes at this point are like late changes in any form of construction. With frame, for example, you can move a window after walls are built, insulated, and closed in, but it's enough of

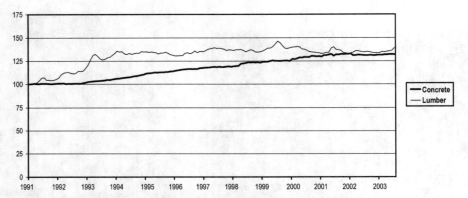

Figure 2-3   Normalized concrete and lumber prices (base at 1991 = 100).
(*Data from Bureau of Labor Statistics, U.S. Department of Labor.*)

a hassle and expense that no one wants to pay for it. So you try to catch any changes before it gets too late, and you resist them later.

Price Stability and Accurate Job Cost Estimates

Both the prices and the quantities of the materials used in ICF construction have been fairly predictable. If you lived through the lumber price spikes of the 1990s, you know what kind of havoc price movements can wreak on your budgeting and profits. Fortunately, concrete prices have not moved up and down nearly as much (see Fig. 2-3).

In addition, the low amount of waste in ICF construction leads to a more predictable quantity of materials. Typically, new crews lose about 5 percent of their forms to waste. The good ones get that down to 2 percent.

The story with wood is a bit different. A study conducted by the National Association of Homebuilders (NAHB) Research Center found that lumber is one of the biggest types of waste generated on a residential job site. And every piece of wood tossed out is an extra charge for the project. But in addition, there is a significant cost to removal. A separate survey by the same organization estimated the average waste removal cost per site at $500. The problem has become worse as our old-growth forests have disappeared and lumber producers resort to supplying more off-spec lumber. It is common for framing crews to report waste levels of 10 percent.

Business Image

A useful advantage of ICFs is that they can help you create a progressive image for your business. This is good for impressing customers and getting press coverage and good word-of-mouth.

ICFs create a building shell that has advantages over the alternative materials, and you can pull out the literature to prove it. ICFs can contribute to making you an "energy-efficient" or "green" builder, two concepts that are currently important to many buyers. All of these things are worth bringing up to potential customers.

Building with ICFs might also get you free publicity. Depending on where you are, ICFs might be considered new and different. In that case they are a good story for the local paper or television station. If you do get coverage and remember to stress your name so that people look you up, it can produce some good leads. According to Keith, a general contractor building his first ICF house in Massachusetts:

> I called a reporter I know at a local TV station. I left a message that we were building a house out of "concrete and coffee cups" just to get her attention. A week later they had a crew visit the job site and we were on the evening news.

One more thing that enhances your image is the cleanliness of the job site. With less waste it is fairly easy to keep things looking precise and orderly. The foam formwork also has a clean, sharp look to it.

Unfamiliarity

ICFs are unfamiliar to a lot of people. It's usually not much of a problem to get the other trades to understand them and work with them, but if you're in an area that hasn't had any ICF construction in the past, you may have to do some more convincing when it comes to the building department and architects.

Consider the experience of Matt, an ICF contractor in Ohio:

> I was recently on a project in Florida as a consultant. After we had completed stacking the blocks, it was to be inspected the following day (Friday) when we had the concrete scheduled. The inspector showed up at the job site and asked, "What is this?" We explained what it was and how it worked and showed him the engineering. He said, "I don't know what this is. You'll have to get another inspection Monday by someone who has seen it." He turned the project down and left. He didn't even look at the project. He didn't even walk into the yard. We had to cancel the pour and reschedule for Monday. Another inspector showed up on Monday, passed the project 5 minutes later and we poured the walls.

In more and more areas the building inspectors have seen ICFs and don't require anything special of an ICF building. But in some places no one has inspected an ICF job. The inspectors may be uncomfortable with the product and place extra requirements on the project. In residential construction they may ask for an engineer's stamp on the plans, or at least the ICF portion of them. Getting the house engineered may be a sizable extra expense. That's usually not an issue in commercial construction, because most commercial buildings are engineered and stamped, anyway. But in a commercial job the inspector may ask for other extra measures, like an engineer's inspection.

Engineers rarely have any concerns you can't deal with quickly. Structurally, ICFs are simply reinforced concrete, which they understand well. And there are complete sets of tables and engineering documentation to reassure them and show them how to do their calculations.

Architects are mostly an issue in commercial construction. Architects are hired for a small minority of the houses built today, but they are retained to design almost all the commercial buildings. And to use ICFs they have to be comfortable with them.

Chapter 13 gives time-tested ways for educating both building officials and architects about ICFs. In most cases people's resistance need not delay a project greatly, but you may need to be prepared for it.

And bear in mind that the whole situation of product familiarity and getting people comfortable with ICFs is getting a *lot* better and may not even be an issue at all in a few years. The construction community and the public have been bombarded with major advertising and information campaigns for almost a decade, and that will only increase as the industry continues to grow and the ICF companies get bigger budgets. The technical groundwork to support the architects and engineers is now very thorough. As discussed in Chapter 13, ICFs are now covered in the key building codes right alongside wood frame and concrete block and everything else. There are special engineering handbooks available with all the tables and charts and formulas. A lot of the ICF manufacturers have reports from the big model code organizations on their own products, as well as a stack of reports on tests run by major laboratories. The companies also have engineers on staff to talk to local designers and building officials if necessary.

So the problem is truly a familiarity problem—there's plenty of experience with ICFs and plenty of information available on them. At worst, you usually just have to get that information to the doubters.

The Bottom Line

ICFs are a *new venture*. They offer the potential for more business and higher profits. To get there you have to learn some new things and bear some expense. The expense is not high as new trades go, but let's face it—money is money.

So you need to follow the usual business guidelines of evaluating a new venture. You need to form an opinion about what the potential payoff is and what the time and money you need to invest will be. This chapter has given you an overview of factors in the payoff and the cost. As you go through the rest of the book you can fill in more details. You can decide for yourself whether ICFs are for you and how you will get yourself into the business.

3

Market Demand

Overview

Ed, an ICF distributor in Minnesota, explains:

> My experience in Minnesota so far has been that the market is driven almost exclusively by the consumer. Switching traditional stick builders over takes some work. However, I have seen considerable growth in the amount of people requesting solid ICF homes.

Ed's experience matches almost everyone else's—maybe 7 out of 10 ICF buildings were built that way because the owners got enthusiastic about ICFs on their own and went out looking until they found contractors who could build with them. In most of the rest of the cases a contractor proposed the idea to the buyer, and in a very few cases an architect or someone else suggested it.

The national promotional campaigns of the Insulating Concrete Form Association, the Portland Cement Association, and the Cement Association of Canada have raised public awareness of insulating concrete forms. Over the past 7–8 years, local newspapers, radio and TV stations, and magazine articles have picked up the story. Every major home improvement show on television has run a segment on ICFs at least once, and a surprising number of people know about the product from seeing one of those spots. And increasingly, consumers are finding out about ICFs on the Internet. They search and find the same sites about ICFs that contractors visit. They read and learn until they are surprisingly well informed.

Commercial construction with ICFs is a little less owner driven. Although many building owners decide by themselves that they want an ICF building, at least as many are responding to a proposal from a contractor or local distributor. The contractor and distributor see the volume of work that's possible for commercial customers, and organize themselves to make contact with these customers, find out what they're building, and make a formal pitch.

The Big Picture

Ian, an ICF distributor in Texas, tells a familiar tale about his buyers:

> Mostly the people who come to us are high-end home buyers. These people are equivalent to the "nerds that built the computer" in character. They are typically well-educated and are looking for alternative, earth-friendly materials. After the home buyers, it is a toss-up between builders and architects. They are looking for a quality structure, higher quality than what is traditionally out there. The main selling points are the storm/wind resistance and the utility saving features.

These are recurring themes in the ICF market: the wealthier and more educated buyer, the bigger house, and the appeal of energy efficiency and disaster resistance. Ian put it well. But don't be hasty. The market is growing and expanding into new areas. There are different kinds of higher-end buyers cropping up. ICFs are now proving popular with some types of affordable housing projects. They are also becoming popular with some middle-market buyers, especially those who are most sensitive to natural disasters and issues of energy efficiency.

You need to know about these up-and-coming buyers and markets, too. They just might turn out to be the best market for you. Or they might provide you with the extra sales to put your business over the edge to growth and profitability.

Who the Buyer Is

Of course there are people of almost every type and from every walk of life who buy ICF houses. However, if you start counting you find that there are currently three groups that stand out. Likewise, many different types of businesses are buying ICF commercial buildings, but there are some broad patterns there, too.

The young homebuyer

The largest single group of ICF buyers are a bit on the young side. The usual characteristics of the young ICF homebuyer are as follows:

- Around 30 to 40 years old
- Higher income
- Educated
- Looking for a second or third house
- Environmentally conscious
- Expecting to live in the new house for 10 to 15 years
- Well informed about ICFs

When you think about this list, it makes sense. Imagine that you have a person with a little more money than the average (upper income). Say also that the person likes to study things (educated), is already familiar with the typical house and its limitations (looking for a second or third house), and is ready to establish roots a bit more permanently (expects to live in the new house for 10 to 15 years). Say also that the person is concerned not only with getting a better house to live in, but one that seems to make sense for society (environmentally conscious). So, being studious, he or she naturally does some reading and watching about homebuilding.

Now, what is the likely result? This prospective homebuyer comes across the extensive public information about ICFs (well informed about ICFs). And the way things read, ICFs fit the bill—they have savings and durability that makes sense for a long-term house and they are environmentally responsible. They cost a bit more up-front, but this buyer has the income to pay for them.

The old homebuyer

Surprisingly, the next most common homebuyers of ICFs are quite elderly. These people are typically

- Over 50 years old
- Upper-income group
- Looking to build their final home
- Well informed about ICFs

This also makes sense. According to a recent survey by the National Association of Home Builders seniors are looking for low-work, low-maintenance homes that would allow them to travel, socialize, and pursue active lifestyles. Like others, they may look through information on homebuilding and happen to find out about ICFs. An ICF house promises great benefits for living, low energy use, and a durable structure. The slightly higher initial cost is not an obstacle because they have above-average incomes and probably have substantial savings.

The housing official

Another group of buyer of ICF houses that is smaller but growing rapidly is the government or private program for subsidized housing. As mentioned earlier, some municipalities like Lubbock, Texas are specifying ICF construction for their new housing programs for low-income residents. Consider also the example of Habitat for Humanity. This is the private organization that uses partly volunteer labor and donated materials to build reduced-price houses and sell them to qualifying low-income citizens. Several local chapters of Habitat for Humanity now build heavily with ICFs. The Kaw Valley Chapter, located just outside of Kansas City, built 11 of its 18 houses in 2002 out of ICFs, and expects to build most or all out of ICFs in the future. According to Kelly Willoughby, the director of the chapter:

> We tried some ICF construction in 2001. The cost was a little higher, but the staff
> calculated that the energy savings more than made up for that. It leaves the owner

a lower monthly bill to pay… ICFs also produce a real quality house, our volunteers have found them easy to work with, and they've stood up well under some tough conditions.

These officials' decisions to specify ICFs make logical sense. The costs of operating the house are lower, which is particularly important to a low-income owner. The initial price is not as important as it might be in other cases because it is subsidized.

High-wind areas, mid-income owners, and other homebuyers

Before you jump to conclusions, don't assume that the three groups of homebuyers listed above are the *only* interested groups you'll find in your area. Every area is different. You might find a high level of interest among some other groups in your town, and you would be foolish to ignore it just because these people aren't mentioned in some book.

One thing that clearly changes from place to place is interest based on natural disasters. Steve, an ICF distributor in Missouri, noted:

I've supplied ICFs for several houses just in the last year where the buyer wanted them because they stand up well in a big blow. We're in tornado alley, and you can't help think about that. These were people who wouldn't have been real likely buyers if the wind issue weren't out there.

Contractors around the country have noticed that people are much more likely to look to buy an ICF house if they live in a high-wind area. Concerns about the danger of hurricanes is high from Texas, all along the U.S. Gulf Coast states to Florida and on up the Atlantic states through North Carolina or Virginia. Concerns about tornadoes is high from North Texas through Oklahoma, Kansas, Missouri, Illinois, and some of the surrounding states. And these concerns rise after every severe storm that makes the headlines. In high-wind areas interest in ICFs is heightened enough for many people in the lower and middle-income groups to ask for the product more frequently.

A different important factor is that ICF construction costs are gradually declining. As that happens, the prices on the houses will edge down. So it's a safe bet that in the future—perhaps just as you begin to build—more and more buyers in the middle- and lower-income brackets will be buying too.

Keep an open mind as you talk to people and research the market in your area. Do any of them seem to be buying ICF houses that don't fit the mold? Why are they buying them? Are there likely to be more like them in the future? You don't want to base a business on one or two customers who think differently from everyone else. But if you can spot a new type of buyer in your area, you may be in a great position to sell to them before everyone else does.

Commercial buyers

It is harder to put commercial buyers of ICF buildings into categories. Commercial buildings come in a very wide variety of types and uses to start with.

And the owners who specify ICFs don't fall under one type of company or building. Businesses of several, very different types request ICFs.

There do seem to be a few common threads, however. Eighty times out of 100, the building holds people for some purpose or another. Out of the 20 times it doesn't, in 19 of them it requires some sort of indoor climate control. And in almost every case, the buyer will own and operate the building—he won't be a developer who simply resells the space or leases it to someone else who has to take responsibility for it.

All of this makes sense when you think about it. Many of the benefits of ICFs—safety, comfort, sound reduction—are benefits to people. Crates and supplies don't much care. If the building houses, say, vegetables that must be maintained at a specific temperature and pressure, then the owners may also be interested in the energy and climate-control benefits of ICFs.

Common ICF commercial projects include

- Apartments, hotels, stores, offices, churches, because giving people a pleasant environment and saving energy is important.

- Movie theaters, because the indoor environment, energy bills, and sound reduction are all important.

- Wineries, freezer and controlled-storage facilities, because control can be more precise, and energy bills are lower.

Conversely, you don't find many ICFs in projects like standard (not climate controlled) warehouses and distribution centers. There the energy and environmental control benefits may not be as important. They also have not sold quite as widely for use in things like prisons and factories. The walls of these types of buildings often endure heavy physical abuse. Like any wall system ICFs can withstand this if they are finished with appropriate materials. But economics usually dictates going with a wall system that has a hard face, like block or precast or tilt-up concrete.

As for ownership, it clearly makes a difference whether the buyer, who decides what the walls are made of, will be occupying or paying for it in the future. If the original buyer just resells it or leases it out to someone else who will have to live in it and pay the bills, he might not put much weight on long-term considerations like comfort and durability and energy costs. Those are someone else's problem. Apartment buildings are a good example. You get a lot more interest from an owner who will be responsible for paying the heating and cooling bills than you will from an owner who separately meters the units and has tenants paying for the fuel.

What the Buyer Is Looking For

Homes

The characteristics of ICFs that are most important to ICF homebuyers are as follows:

1. Energy savings
2. Wind resistance

They mention these things most often when they explain their interests to their builders and they spend more time talking about these than any other feature of the product.

According to John, an ICF contractor in Colorado:

> People ask about all kinds of little things, like how do you hang a picture and how do the contractors attach siding. But fuel costs and how they survive high winds are the ones they talk about the most and those are the things that seems like it makes most people want to buy the ICFs. I always mention a lot of the other benefits, too, like comfort and sound and insurance savings. But those are more like the icing on the cake. You don't talk about them too long because they are more interested in the other things and you don't want to distract them or look like you're dodging their questions on the things that are most important to them.

As you would expect, energy savings are about equally important to buyers everywhere, but interest in wind resistance is much more regional. In areas with low winds the buyers are usually interested more in energy, and wind is one of the "other" benefits that gets discussed a little. Along the coast or in tornado alley, wind may be the main interest and energy is second.

But be careful. The things that owners of an ICF house mention as *the things they like best about the house after they are in it* are very, very different. A survey in 1997 asked over 70 ICF homeowners from all over the United States and Canada what things they liked best about living in their house. They got to mention as many things as they wanted, and most people mentioned two or three. The most-mentioned benefits were as follows:

Benefit	Mentioning (%)
1. Comfort	81
2. Sound reduction	65
3. Energy efficiency	43
4. Strength (wind resistance, lack of vibration, etc.)	31

The explanations people gave for their views were very revealing. No one was disappointed by their energy savings or the strength of their houses. In fact, a lot of them bragged about how much lower their fuel bill was than the neighbors'. But they raved about the comfort and quiet. Apparently, before they lived in these houses they did not believe that these things would be much different from their old house, or else they didn't think they would find them so important. But when they actually got to experience the comfort and quiet for themselves they were delighted. In contrast, the energy savings were something they didn't feel every day, and most of them had not been through severe wind storms, so those things were a little farther back in their minds.

Peter, an ICF homeowner in New England, said:

> Building our house out of ICFs was my idea, and my wife kind of thought I was exaggerating all the great things about them. Well, after we'd lived in the house a couple of months we went to dinner at her parents' house. It was November and kind of cold. It was a little nippy inside, too. After a while the heat came on, and in a few minutes we were hot and we took our sweaters off. Then the place cooled down, and in 15 or 20 minutes we were chilly and put the sweaters back on. Then the heat clunked on again and we got too hot all over again. This went on and on, and on the ride home my wife said "Wow, I never realized how much better our house is." We never felt any ups and downs like that in our new place.

Karen, an ICF home owner in Nebraska, said:

> It just gives you a *feeling* of being very, well, *solid*. That's the best I can describe it. Just being surrounded by something very solid.

But make sure you do *not* confuse what your homeowners say they like after they live in their house, with what your homebuyers want to hear about before they buy the house. The buyers want to talk about energy and strength. They will probably be less interested in hearing about comfort and sound. You can be quietly satisfied knowing that the house they get will exceed their expectations.

Commercial

Not surprisingly, when commercial building owners consider using ICFs they follow a line of thought that looks more like a business analysis. They usually lay out the costs and the benefits that are important to them and frequently ask for hard numbers to measure everything.

In fact, one of the reasons ICFs appear to be doing so well in the commercial market is that in many projects they actually do what the buyer wants less expensively than the alternative wall systems do. The competition often requires lots of additional work and materials to meet the more stringent requirements of the commercial project. So the cost of the other systems goes up sharply, and ICFs may actually end up being less expensive.

In commercial construction, the building loads are usually higher, so frame construction would have to be beefed up and becomes more expensive. The concrete of an ICF looks good by comparison because it is usually strong enough or it can be made strong enough with just the addition of some rebar. For many businesses, energy bills are a major part of the budget, and the company gives full dollar weight to the future fuel savings that ICFs can provide. Things like sound reduction are extremely important to businesses like movie theaters and upscale hotels, and they have exact minimum specs for them. Even things that are hard to quantify, like indoor comfort, many companies know are important for their customers or their employees, and so they take them specifically into account.

But the flip side of all this is that it is impossible to give a single rule for what the commercial ICF buyer wants. It's different for any one buyer.

One thing you *can* do is make a reasonable guess about who will be interested in ICFs and why. That's why it makes sense that a movie theater owner might be interested (sound, energy efficiency, comfort), and a winery owner might be interested (precise environmental control), but the owner of a simple warehouse (none of the above) might not.

4

ICF Products

Overview

At last count there were about 50 different brands of insulating concrete forms sold in North America. But there's only a handful of major ones, and among these there are a lot of similarities. The parts and terms are mostly the same. Contractors who learn on one system can usually pick up another one without much trouble.

There are also a lot of useful related products available now. Most of the form companies also sell special forms for special purposes. There are new tools and materials and equipment that make common jobs like cutting, gluing, and aligning the forms faster, easier, and cleaner.

This chapter covers all the new products you need to know about to get started with ICF construction. You can track down sellers of these on *www.icfweb.com*, the independent web site, or by going to "Member Search" on *www.forms.org*, the web site of the Insulating Concrete Form Association, since most of the major players are ICFA members.

Wall Forms

Insulating concrete forms are sold by dozens of companies. Almost all the major ones are members of the Insulating Concrete Form Association. You can find them listed in the directory of the ICFA's site, *www.forms.org*, under "Primary Members."

Most ICF companies offer one **system**. A system is a set of forms that fit together to build walls. Forms from one system won't always fit with forms from another system. Some of the companies now offer more than one system to give the contractor a choice. The forms come in two types—**blocks** and **planks**.

Blocks

Most blocks are 16 in high and 48 in long, but some companies sell bigger or smaller ones. Every supplier has blocks of different widths so you can build

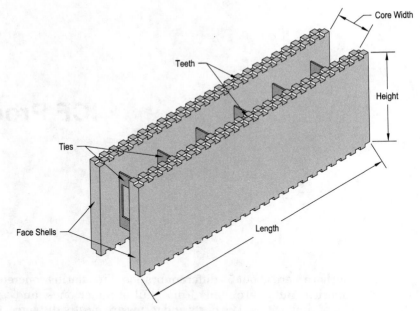

Figure 4-1   A typical ICF block.

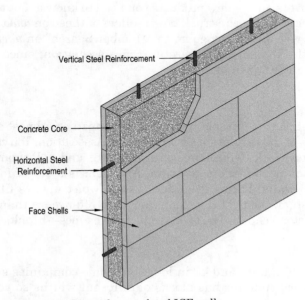

Figure 4-2   Parts of a completed ICF wall.

Figure 4-3 Cutaway views of flat (left), waffle (center), and screen (right) concrete walls. (*Courtesy of Portland Cement Association.*)

walls of different thicknesses (see Fig. 4-1). The wider block has a bigger cavity, so the finished wall has a thicker concrete section and is stronger (see Fig. 4-2).

The two layers of foam that make up most of the block are called **face shells**. Almost all blocks have teeth or ridges along the edges of the face shells to connect the blocks together, a lot like children's building blocks.

Within the blocks there are three subtypes—**flat blocks**, **waffle blocks**, and **screen blocks**. The shapes of the inside surface of these forms are different, so the shape of the concrete cast inside turns out different, too (see Fig. 4-3).

Flat blocks have face shells that have the same thickness everywhere—each face shell is a flat piece of foam. The face shells run 2 to $2^3/_4$ in thick, depending on the system. After the concrete is cast inside, it creates a flat, solid concrete wall.

Holding the face shells a constant distance apart are crosspieces called **ties** (see Fig. 4-4). Ties are made of plastic in some systems and steel in other systems. The ends of the ties are molded into the foam. If you ripped a tie out you would see that from the top it has an "I" profile. Each end of the tie has a flat strip called a **tie end** or **furring strip**. The tie ends are the top and bottom sections of the "I". They keep the tie securely embedded in the foam. They also give the crews something to attach things to. This is useful for attaching such items as temporary bracing, sheetrock, and siding. A coarse-threaded drywall screw sunk into a typical tie end has a pullout of over 100 lb. Tie ends run about $1^1/_4$ to 2 in wide. The long midsection of the "I" is called the **web** of the tie. To make a wider block the manufacturer uses a tie with a longer web. A lot of tie webs now have slots or snap-in holes to hold horizontal rebar.

Waffle blocks have face shells that are not flat. They are "wavy" on their inside surface. If you stripped off the foam of a waffle wall, you would see that the concrete has a shape like a breakfast waffle standing on its edge. There are thick vertical and horizontal ribs of concrete with thinner, square sections of concrete in between. The vertical ribs are sometimes called **posts**, and the horizontal ribs called **beams**. The same kinds of ties are used in waffle blocks as in flat blocks (see Figs. 4-5 and 4-6).

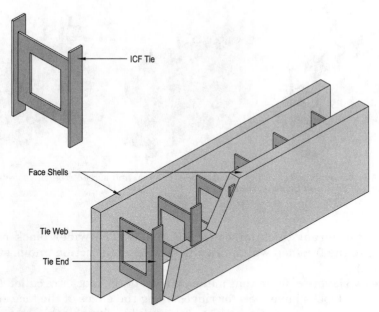

ICF Tie

Face Shells

Tie Web

Tie End

Figure 4-4 The parts of a typical tie.

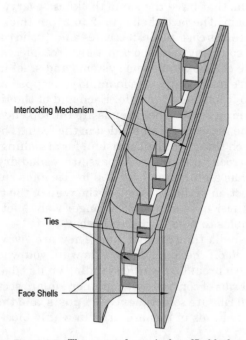

Interlocking Mechanism

Ties

Face Shells

Figure 4-5 The parts of a typical waffle block.

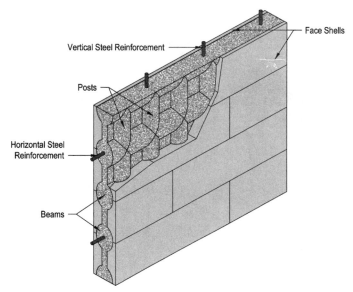

Figure 4-6 Cutaway view of a typical waffle wall.

Waffle blocks save on the amount of concrete needed to fill the forms but require a bit more attention to line things up correctly.

Screen blocks are all one molded piece of foam, including the crosspieces (see Fig. 4-7). These are not usually called ties anymore, but simply webs, because they are really just an extension of the face shells and there is no "tie end" to

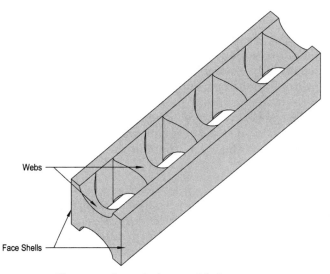

Figure 4-7 The parts of a typical screen block.

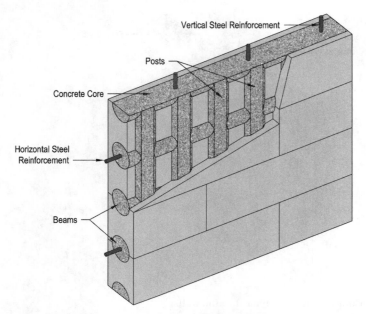

Figure 4-8   Cutaway view of a typical screen wall.

them. If you stripped the foam off the concrete in a screen block wall, it would look like a window screen with thickened wires and small holes in between. The holes are where the foam webs were. Fastening things to a screen wall is done with different methods—usually concrete connectors or threaded rods through the webs for heavy objects, and glue for light objects (see Fig. 4-8).

Screen blocks require even less concrete in the wall than waffle blocks and the blocks themselves may be less expensive. But some buyers prefer to have ties for attachment and an unbroken layer of concrete in the core.

Most blocks come fully assembled from the manufacturer. But also popular are **field-assembled blocks** (see Fig. 4-9). These may save on shipping and storage, but require a little more work in the field. The contractor gets a stack of face shells, and a separate package of tie webs. The face shells have tie ends molded into them, with some kind of slot or nub to snap the webs to. So the crew snaps the blocks together, then puts them in the wall.

Planks

Plank systems work in some ways more like conventional concrete forms and less like building blocks (see Fig. 4-10). The **planks** of the system are pieces of foam that are usually 8 feet long, 8 to 12 in wide, and about 2 in thick. They usually have slots or grooves along their edges. Separate **ties** or **rails** slide into the slots/grooves. The crew builds the wall by putting a pair of planks in

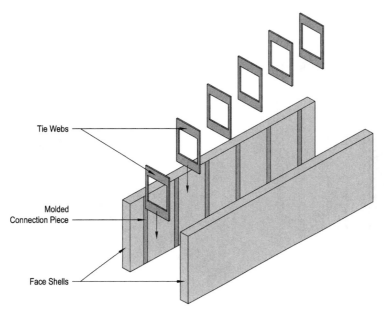

Figure 4-9 The parts of a typical field-assembled block.

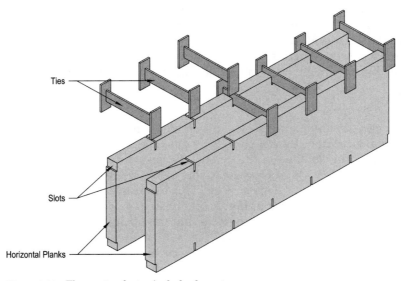

Figure 4-10 The parts of a typical plank system.

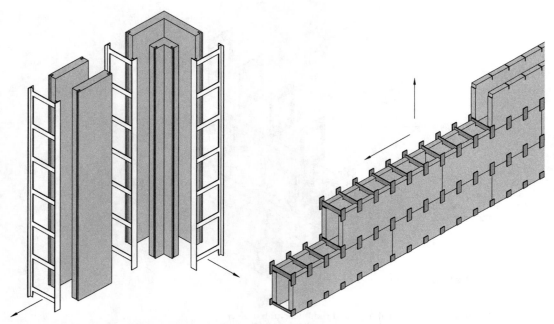

Figure 4-11   The wall layout of horizontal (right) and vertical (left) plank forms.

position (one for the inside and one for the outside of the wall), then sliding in the ties or a rail along their edges. Plank systems also save on shipping and storage, with some extra steps involved in assembly.

Plank systems also come in two varieties: **horizontal** and **vertical** (see Fig. 4-11). Horizontal planks are set into the wall horizontally in about the same pattern as blocks. Vertical planks are cut to the full wall height, so they stand up side by side in the wall, more like conventional concrete formwork.

Speciality forms

You can cut and paste blocks and planks to make the formwork for almost any special situation. But for tasks that get repeated a lot, the manufacturers have special forms molded in the factory. These can save a lot of field labor.

Almost every system has some sort of pre-made **corner** available. For blocks this is usually a 90° dogleg block. This can come in a **left** and a **right** version, or one reversible version. Plank systems have pre-cut 90° corner plank pieces, or sometimes a special **corner tie** that the straight planks fit into to butt them together and form a corner (see Fig. 4-12).

A few systems even offer pre-made corner parts for 45° and 60° angles.

Special parts are also available to create a brick ledge. Many block systems include a special **brick ledge block** that is stacked directly on the regular blocks

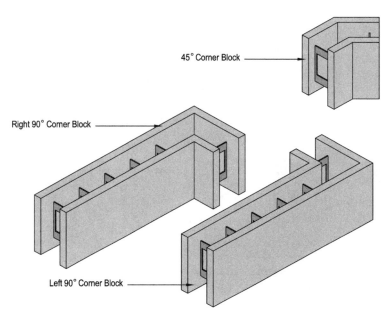

Figure 4-12 Typical corner blocks.

(see Fig. 4-13). It flares out on one side to form the ledge. Accessory suppliers also offer general-purpose **brick ledge forms** that can be attached to any ICF wall.

Most systems have only one size of block or plank. But a few offer a larger and/or a smaller version of their standard form. The larger ones may help stack up wall area faster. The smaller ones let the crew create different wall heights or lengths without cutting forms.

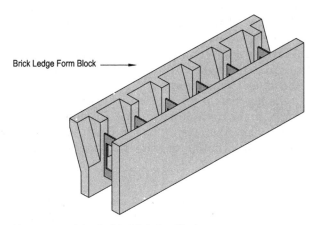

Figure 4-13 A typical brick ledge block.

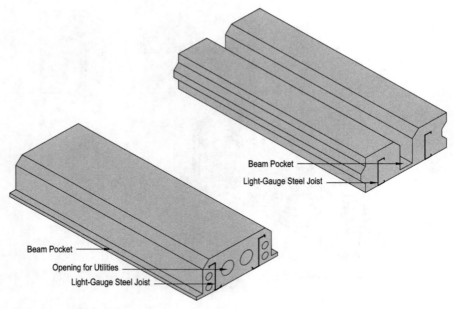

Figure 4-14 ICF deck forms.

Deck Forms

In the last few years, special ICF deck systems have appeared in North America and gained popularity. Several of the ICF companies sell them. These create an insulated, reinforced concrete floor or roof that ties into the walls. They can be used on any ICF wall.

ICF deck forms are made of foam with light-gauge steel joists molded in (see Fig. 4-14). The crew sets them up on the walls at the top of the story to create a floor deck or a roof. They are provided in almost any length, and can be cut down in the field. They have beam pockets along their length about every two feet on center. They are available with beam pockets of different depths to create decks of different strengths and spans. Rebar and concrete go on top to create a floor that ties in to the walls. Utilities can be cut into the foam and wallboard fastened to the joists from below.

The use of ICF decks for roofs is partly limited because it is difficult to pour concrete with a pitch of more than 4 in 12. That is enough for most warm climates. In cold climates, the engineering has to be checked to make sure that the low-pitched roof is strong enough to carry the local snow load.

Foam

ICF foams are made from plastic. The manufacturing process fills the plastic material with so many tiny beads of air (or some other gas) that its properties change. It becomes light and workable and insulates extremely well.

The foam used in ICF formwork is usually **expanded polystyrene**. This is abbreviated EPS. Contractors sometime refer to this as "beadboard" because it is made by fusing together a lot of tiny foam beads in a special heating and expansion process.

EPS comes in different **densities**. The density tells how much of the foam is plastic, versus air. Denser EPS weighs more and is a little stronger, more expensive, and very slightly more insulating. The EPS used in ICFs usually has a density of 1.5 or 2.0 pounds per cubic foot. Either one is adequate for the job with correctly designed formwork.

Extruded polystyrene (XPS) is also available with some plank systems. The XPS used on ICFs is virtually always a density of 2.0 pounds per square foot (psf). As a rule it is a bit stronger and has a higher insulating value than EPS, so some buyers may prefer to pay extra for it.

Some suppliers now offer forms made of **borated** EPS. This EPS contains fine borate powder, which is a proven termite killer but considered safe for humans and animals. Using forms with borated foam is one of the methods of deterring termites that is accepted by major building codes for use in areas of high termite risk. They are rarely used in other areas because there the risk of termites is considered lower and the codes do not require any special termite precautions on ICF buildings.

Tools and Accessories

Most of the work of ICF installation can be done with standard tools and parts that contractors already carry. But as the ICF industry has grown, inventors and suppliers have produced a large set of special devices to make construction faster, easier, and more reliable. They are available from regular ICF suppliers and several mail order construction supply companies.

The forms are cut with just about any tool used to cut lumber. Forms with steel ties may limit the tools you use a bit, but standard power saws and many hand saws can all be made to work without much trouble.

One popular new tool for cutting foam is the **hot knife** (see Fig. 4-15). This is an electric hand tool with a metal blade that heats up. It makes especially clean, precise cuts. Suppliers now have a wide variety of models and accessories to perform different cutting and shaping operations in the formwork.

It is often useful to stick pieces of formwork back together or reinforce a joint. This is done mostly with glues and tapes. Just about any construction adhesive that doesn't dissolve foam will do the job. But regular ICF users usually buy **low-expanding foam adhesive**. It not only adheres well, it also fills small gaps. This means that cuts do not have to be completely precise because the foam will fill between the pieces. In addition, damaged spots in the block are easy to repair by filling them with the adhesive. The foam adhesive comes in pressurized cans that fit into an applicator gun (see Fig. 4-16). Most contractors find the setup to be worth the cost.

When they use tape, most people find that ordinary fiberglass-reinforced packaging tape works well. It sticks and is strong enough to hold up against a lot of concrete pressure.

Figure 4-15 A hot knife. (*Courtesy of Reward Wall.*)

Figure 4-16 Low-expanding foam adhesive and applicator gun. (*Courtesy of Arxx Building Products.*)

Figure 4-17 A buck made of a special plastic channel going into the wall. (*Courtesy of American Polysteel.*)

Plastic bucks

Bucks are the subframes ICF installers put in the wall at openings (see Fig. 4-17). They hold the concrete in the wall. When it's time to insert the doors and windows, the crew can fasten them to the bucks.

Originally almost all bucks were made of conventional lumber. But now plastic channel designed specifically for this job is widely available. These products may come with special corner pieces that snap the sides together into a precise 90° angle. The material itself is more expensive than lumber, but most contractors claim the plastic products reduce assembly labor. They also reduce waste because even short scraps can be pieced back together and used.

Bracing/Scaffolding Systems

ICF form walls require something to hold them precisely straight and plumb through the concrete pour. The workers also require something to stand on when they construct the upper levels of the wall. These needs can be met with various homemade arrangements of 2 × 4s, but most contractors have shifted to renting or buying pre-built **bracing/scaffolding systems** (see Fig. 4-18). These also go by the name of **wall alignment systems**.

Bracing/scaffolding systems are usually made of steel or aluminum parts. A few consist of standard wood or steel studs held together with special premanufactured brackets. In the typical system, once the wall is at eye level the crew

Figure 4-18 A typical bracing/scaffolding system. (*Courtesy of Insulating Concrete Form Association.*)

erects a post along the form wall about every 6 feet. Workers attach each post to the wall to hold it upright. This is usually done with a screw or two into the ties. Connecting scaffolding posts to a wall of foam may sound risky but once the bracing/scaffolding is fully in place it's amazingly sturdy. An adjustable kicker runs from each post down to the ground or floor inside. The posts have brackets to hold walk boards, toe kicks, and hand rails. The kickers have a turnbuckle or other adjustable extension to set the posts to precise plumb.

Scaffolding generally has to meet special OSHA standards. Most commercially available bracing/scaffolding systems meet them and the manufacturers can provide the paperwork to prove it.

Concrete and Steel

Overview

ICF forms are filled with **reinforced concrete**. This is concrete with steel reinforcing bars (commonly called **rebar**) cast inside. The concrete gives the wall strength against compression (called **compressive strength**) so it can safely bear the weight of things pressing down on it. The rebar gives the wall strength to resist things that try to bend it or lift it up (called **tensile strength**).

Most ICF installers were trained as carpenters. The biggest single thing most carpenters have to learn to use ICFs properly is how to specify and handle concrete. This chapter gives an overview of the factual information you need in these areas to understand ICF construction. But it cannot give you the "feel" of handling concrete. That you'll get when you're trained.

Handling and installing rebar involves learning a few more things. Most of these are also presented here.

Ingredients

It's easy to tell people who really know something about concrete from those who don't. Just find out whether they understand two key facts:

- Concrete is *not* the same as cement.
- Concrete does *not* harden by "drying out."

In construction, the term cement almost always refers to **portland cement**. This is a gray powder made by grinding up and heat treating certain minerals. Portland cement has the almost magical property that it hardens into a rock-like mass when mixed with water.

Concrete almost always refers to a mixture of portland cement, water, and **aggregates** (sand and stones). Concrete also hardens into a rock-like mass, because with the water present, the cement hardens up and binds all the aggregates together.

The hardening is a chemical reaction that occurs when the cement and water are mixed. Engineers call it **hydration**. (Contractors usually call it **curing**.) The cement does *not* harden when the water evaporates. In fact, it continues hardening more and more and more as long as there is still some water present. So if cement or concrete dries out quickly, it will actually be much weaker than if it remains moist. Hydration is like the hardening of a two-part epoxy. The two ingredients separately have no strength. But when they mix they create a chemical reaction that makes them harden into a mass. Hydration is *not* like white glue. White glue starts out wet and hardens as it dries out.

We build a lot of things out of concrete for good reason. It is strong and very durable. Once it hardens, it is almost totally unaffected by water or the elements. Insects and vermin can't eat it or get through it, and it doesn't burn. In fact, heat has no significant effect on it until you hit those very rare fires that get into the thousands of degrees.

Properties

For the sake of the final building, the compressive strength is far and away the most important property of the concrete. Compressive strength is how much pressure the material can take before it shatters. And since the walls have to hold the rest of the building up, this is critical. The ready-mix supplier can vary the strength of the concrete by adjusting the amount of cement and water in it. Most buildings call for concrete with a compressive strength of either 2500 or 3000 pounds per square inch (psi).

The compressive strength is usually the only concrete property that engineers or the building code specify. But the ICF contractor is also concerned about the cost and the **flow**. Flow refers to the ease with which the concrete flows around rebar and ties and into all the corners and crevices in the formwork. Some people say **flowability** or **workability** to mean the same thing.

The most widespread measure of how well concrete flows is called **slump**. It's measured with a **slump test** (see Fig. 5-1). In the test, someone fills a special 12-in high steel cone with concrete according to a specific set of rules, then pulls

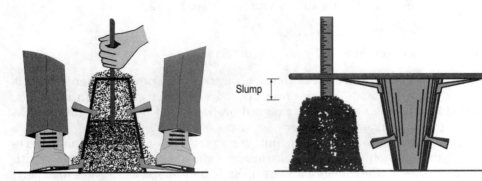

Slump

Figure 5-1   A slump test.

off the cone. It's something like filling a bucket with wet sand, then turning it over and taking off the bucket to make a sand castle. After removing the cone, you measure how far down the mound of concrete slumps from 12 in. That distance is the slump. Typically ICF contractors use concrete with a slump of about 5 in. Much more than that and the concrete puts high pressure on the forms and is hard to control. However, some experienced contractors have come up with concrete mixes that have a slump of up to 7 in that they can pour efficiently. The advantage of high slump is that the concrete fills the forms quickly and thoroughly.

There is one very important influence on strength that the contractor has to watch out for and that is the amount of water in the mix. All too often contractors add extra water, and this can harm the properties of the concrete.

Only small amounts of water are necessary to make the cement in the concrete harden. Testing shows that about 3 lb of water for every 10 lb of cement is enough. A mix with these ratios of materials is often said to have a **water-cement ratio** of 0.3 (see Fig. 5-2). However, concrete with such a low ratio is almost never used because it is far too stiff to work with. So more water is added to increase the flow of the concrete. But, on the other hand, the more the water added beyond this amount, the weaker the final concrete will be. Experience shows that water-cement ratios of about 0.4 to 0.5 make the concrete flow well enough for most purposes and leave it with adequate strength. The ready-mix supplier adjusts the water, cement, and the other ingredients as needed to make sure the strength of the mix meets specifications.

The reason the contractor has to watch out is that it can be very tempting to add excess water in the field. The concrete may arrive at the job site a bit stiffer than the workers would prefer. So they order the addition of water. Adding a little

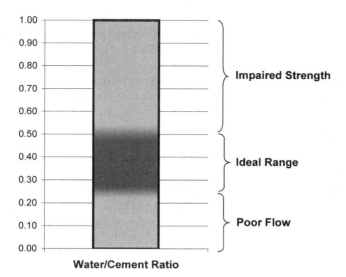

Figure 5-2 Water-cement ratios.

water isn't necessarily bad. The ready-mix driver is supposed to know how much water is in the mix and how much more can be added without weakening the concrete below specifications. But on occasions he doesn't or the contractor disregards his warnings. Needless to say, it is dangerous to use concrete that is weaker than the requirements call for, and the contractor is liable for any bad results. Adding just 10 gal of water to a yard of concrete can reduce its strength by 10 percent.

There are a lot of other ways to improve the flow of the concrete without adding water. These include such things as using smaller aggregates and adding special ingredients called **plasticizers**. Most of these measures increase the cost of the concrete. However, most ready-mix suppliers can advise you on the best ways to balance out the strength, flow, and cost. Just take the time to sit down and explain your needs.

And be sure you know exactly what admixtures will do before you use them. For example, most plasticizers make concrete flow more freely for an hour or so. After that, they actually *accelerate* the curing—the concrete stiffens up fast. This can get you in trouble if you don't plan for it.

A lot of the things crews do in handling concrete are to control the amount of water after it leaves the truck. The concrete hardens rapidly for the first couple of days. It continues to cure over time, but more and more gradually. However, if the water in the mix is allowed to escape early, curing slows or stops. The resulting concrete can again be too weak.

This state of affairs is the reason for a lot of the do's and don'ts of concrete. It must not be allowed to dry out, so after pouring in dry, hot-climate crews sometimes cover it with something like plastic sheet. Formwork is supposed to be left up for days or weeks. If it's removed too early, the water escapes and the curing may slow down so much that the concrete never reaches its designed strength (see Fig. 5-3). The concrete must be kept above freezing, or ice will form and rob the mix of the needed water.

Conversely, crews must be careful to keep extra water out of the concrete. It should not be poured during heavy rain, and water that gets in the forms needs to be removed before putting concrete in.

Fortunately, ICF forms are a pretty ideal container for curing concrete. The insulation keeps the concrete warm, even through a few days of subfreezing weather. The forms don't readily let water out or in, except maybe at the top, and that is easy to cover. You never strip the forms, so you don't have to be worried that you are doing it too early.

The other important part of how the concrete is handled is **consolidation**. This is shaking or vibrating of the concrete in the forms to get air pockets out of it. This is important because air pockets, if left in, create weak spots in the wall. It's also important because air pockets around the rebar reduce the concrete's "grip" on the bar. That is critical because the strength of the wall also depends on the tight connection of the concrete and the rebar.

There are several alternative methods of consolidating the concrete that you learn about from your installation manuals and courses. When air pockets do remain in the concrete, once the concrete cures they become empty spaces in the

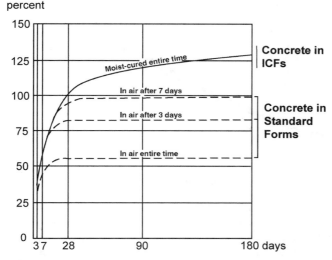

Figure 5-3 The strength concrete achieves (as a percentage of design strength) is higher the longer the forms are left in place. (*Courtesy of Portland Cement Association.*)

wall. These are called **voids**. Voids not only reduce the strength of the wall. They are also unnerving to anyone who happens to tap on the formwork and hears a hollow sound. It's best to avoid them by doing good consolidation. But, as explained later, when voids do occur they are not hard to fix.

Rebar comes in different sizes and strengths. The plans or other construction tables tell what **diameter** of rebar to use. **Number 3** rebar means bar that is ³/₈ in in diameter. Number 4 means ⁴/₈ in (equal to ¹/₂ in), number 5 means ⁵/₈ in, etc. Most rebar in ICF house construction is number 4, with an occasional number 5 or number 3. Commercial construction often uses larger bars because of the heavier building loads.

The length of the bars is usually up to the contractor, and suppliers cut them to any length. What length you choose to use is mostly a matter of work efficiency and convenience.

Rebar is also classified according to its **tensile strength**. This is how much pulling it can take without snapping apart. Rebar used in ICFs usually has a tensile strength of 40,000 pounds per square inch (psi), which means that's the amount of pulling it can bear. This is also called **schedule 40** rebar. Sometimes requirements will call for schedule 60 rebar, which has a tensile strength of 60,000 psi. The two look alike, so the contractor has to be careful to keep them straight.

Tools

There are also a few special tools for handling concrete and rebar. They come from general concrete construction, but some ICF supply companies carry them so the ICF contractor doesn't have to go elsewhere.

Figure 5-4 Vibrator. (*Courtesy of Makita U.S.A., Inc.*)

The **vibrator** is an electric tool, handheld, with a long vibrating shaft (see Fig. 5-4). After the concrete is in place in the forms a worker inserts the shaft in a very specific pattern to consolidate the concrete. When the vibrator is in the concrete, you can see air bubbles rise to the surface. There are other ways to consolidate, but this is considered the most thorough.

It is also useful to have a few **trowels** on site to strike the concrete level at the top of a wall. Some people also carry a **slump cone** so they can do their own testing of the concrete slump.

For leveling concrete floor and roof decks there are several specialized tools and machines that the concrete crew uses. This work is almost always left to a crew that is separate from the ICF installers, so this book tells how to find and coordinate with those people, not how to do the job yourself.

The most used tool for rebar is the **cutter-bender** (see Fig. 5-5). This is a large lever that has blades to cut rebar like a scissors, and holders so that you can bend it to go around corners.

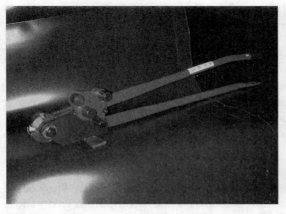 

Figure 5-5 Rebar cutter bender (left) and in use (right). (*Courtesy of Arxx Building Products.*)

Figure 5-6 Tie wires and tying tool.

The rebar also has to be secured in place so it doesn't shift during the pour. Many brands of ICFs now have notches or fingers in their tie webs for you to snap the rebar into. An alternative is to bind the bar to the ties with **tie wires,** a standard industry product that comes in prepackaged bundles. A special **tying tool** is available to speed the job and save your fingers (see Fig. 5-6). Another option that has become popular is common plastic **cable ties**, which also can be applied with their own hand tool, called a **cable gun**.

6

The Construction Process

Overview

ICF contractors have invented thousands of tricks to make construction faster, easier, more precise, or just better suited to local practices and conditions. But everybody has to walk before running. Most people start with the methods that are easiest to learn and easiest to carry out error-free. It's kind of like getting driving directions in an unfamiliar town. The person who gives them to you has you stick to the Interstate as long as possible even if the total distance is longer, because that route involves fewer turns and is better marked.

When you finish this chapter you should know the essentials of building ICF walls and decks using the most common and understandable methods. That's enough to get a feel for what you'll have to do to switch to ICF construction. It's also enough to understand the rest of this book. When you start building, your training will teach you some alternative methods and useful tricks. You'll pick up even more from other contractors and your own ideas. For now, just remember that for every construction step described here there are half a dozen alternatives that are good for different circumstances and styles of operation.

Improving

History shows that ICFs have a short learning curve. According to Paul, an ICF contractor in Florida:

> I've built 20–25 houses and I'm still learning. But I was pretty much up to full speed by the fourth house. My labor time on the second house dropped by maybe 15 percent compared to the first. The third house took about 25 percent less time than the first, and the fourth house took maybe 30 percent less.

Paul's experience is pretty typical. For most crews, the fifth project they build will take 30 to 50 percent fewer labor hours for the ICF walls than the first one did.

This is one reason a lot of people recommend starting ICF construction with something simple. A rectangular basement with a couple of windows is a good starter project. If you're in an area without basements, something like a utility building will do. In a simple box, there are few cuts and corners. There isn't much to mess up. And it doesn't usually matter if the results aren't picture because most of the structure will be underground anyway. Cost overruns or cost underestimates are usually bearable because actual total cost is low anyway.

But even on a simple project the contractor learns most of the skills needed to build successfully. Once he goes on to more complex structures, he has fewer new things to learn and the job is likely to go pretty smoothly.

What Parts Are Built of ICFs

Originally ICFs were mostly used for the basement walls of homes, with wood-frame walls built on top for the above-grade structure. The ICFs were a direct replacement for traditional foundation walls. Their advantage was that they provided a well-insulated basement that was ready for finishing. Projects like this are still done today, and they offer the contractor a good start for learning ICFs.

But today most buyers of ICF houses ask for them in all the exterior walls, including above grade. In the South the ICF walls are usually built off of a slab, the same as frame walls. In moderate climates they are used to construct the stem wall and continue up to the roof. In the North they are used for the basement to the roof.

Something like 30 percent of all ICFs are now used to construct commercial buildings. Most of these are small or midsize structures, up to about 3 to 4 stories, and as high as about 40 ft. Only a few things in the construction methods are different from building houses. Because greater structural strength is usually required, the forms are generally wider to create a thicker wall with more concrete. The plans also call for more rebar. The things that get attached to the commercial building, like floors and roofs, will usually be heavier. So the ICF contractor has to embed heavier steel connectors or weld plates in the walls.

In the last few years a lot of heavier assemblies have been creeping into ICF houses, too. Some large houses have a few *interior* walls built out of the forms. Usually these walls serve as bearing walls for something heavy above. But sometimes they are intended to isolate different parts of the house. Some owners want insulated concrete walls around a media room to contain noise. Some want to separate off a guest space or divide different heating/cooling zones. As a rule, interior ICF walls have to rest on a foundation—a footing or bearing slab. Upper-story ICF walls are built directly on top of lower-story ones.

In recent years, new ICF deck systems have become popular for floors and roofs. These are foam forms used to make a concrete deck, instead of conventional steel or wood forms. Just as with the wall ICFs, the foam stays in place after the concrete cures. ICF decks are more expensive than frame decks. But, according to Peter, an ICF contractor in Florida:

We use foam deck forms for various reasons. You get an insulated concrete floor that is ready for tiling, and is sound and termite resistant. You can do a flat or inclined roof that is almost hurricane- and tornado-proof. Clear spans can be 35 feet or more.

Concrete floors are also attractive because they're very rigid, with almost no detectable flex or vibration. Adding an insulated concrete roof to the insulated concrete walls makes a total building shell that is even more energy-efficient. At this point maybe 5 percent of all buyers are willing to pay the extra for these decks, but that's growing.

A last option that is popular in some areas is constructing an ICF safe room inside a frame building. This is a relatively quick, inexpensive way to give the occupants a measure of safety from strong winds. In the case of a tornado or hurricane they can run into the safe room, much the way they used to run into underground wind shelters.

Design

Architecture

Almost any building design built with frame can be readily built with ICFs. In fact, a lot of the designs used are stock frame plans that the builder adapts. The easiest way to do this is to align the inside of the ICF walls with the interior of the frame walls in the plans (see Fig. 6-1). This way, everything on the inside stays the same. Because the ICF walls are thicker, there will be a small increase in the amount of exterior finish materials.

Most contractors who use this method on homes don't even bother drawing up new plans. They can simply line things up in the field as they go. Commercial construction usually requires that changes of this type be submitted to the designers and documented.

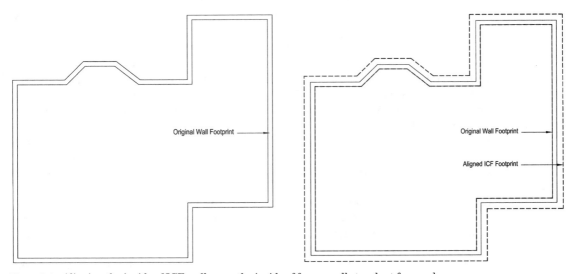

Original Wall Footprint

Original Wall Footprint

Aligned ICF Footprint

Figure 6-1 Aligning the inside of ICF walls over the inside of frame walls to adapt frame plans.

Figure 6-2 Formwork for a curved ICF wall. (*Courtesy of Formtech International Corp.*)

One important set of dimensions that do change when the ICF walls are allowed to grow to the outside are the roof length and width. Roof trusses, for example, might have to be 6 to 12 in longer to extend over the thicker walls. A tad more plywood and shingles will be needed. This is a critical detail to remember when ordering materials.

Some wall features are even a little easier with ICFs than with frame. Curved walls are one (see Fig. 6-2). Odd-shaped windows (like diamonds and circles) may also be easier because it's simple to cut irregular shapes in the foam.

More difficult and costly are **spans** and **projections** (see Fig. 6-3). Spans and projections are places where the walls on an upper story don't line up with the walls below—they are set back or they stick out. A common type of projection is a garrison, where one wall of the second story of a house extends out about 16 in farther than the first story. This can be done with ICFs. However, the method used for frame—build the floor deck to cantilever out over the first story wall—doesn't always work with ICFs. The weight of the wet concrete above flexes the floor out of position.

There are some alternative ways to do spans and projections. But most ICF contractors try to get this kind of feature out of the design or they build the upper-story wall out of frame. The structures that are built with spans and projections are almost always for buyers who want a dramatic architectural building and are willing to pay for the extra work.

If someone else is designing the building, insist on talking to that person early. The designer will probably not know much about ICFs. He or she might

Figure 6-3 Projection built with ICFS.

try to design in spans or projections without realizing they increase time and cost. Or the designer might be avoiding curved walls, assuming they're a lot more difficult. As you become more experienced with ICFs you will probably come up with a list of things that would make construction easier for you, and a list of things you can do that most designers assume you can't. You have to catch the design early—before it's finalized—to get these things considered.

Engineering

Things like the thickness of the concrete in the walls, the compressive strength that the concrete must have, and the diameter, grade, and positioning of the rebar in the wall are all determined by engineering. However, that doesn't mean that an engineer always reviews and stamps all the plans. The concrete and rebar requirements for many smaller buildings have been worked out and put into tables, just as stud and header specs are put in tables for wood frame construction. They are published in many model and regional codes. You plug in the building dimensions, local wind loads, local seismic category, and other numbers related to the building shape and the forces it will be subjected to.

If your code books don't cover ICFs yet, you can go back to the original sources of the engineering. The *Prescriptive Method for Insulating Concrete Forms in Residential Construction, Second Edition*, is the latest engineering book for ICFs. It comes from U.S. government agencies and some trade organizations. It has all the tables and formulas. The sections in the codes about ICFs all come from the *Prescriptive Method*. The *International Residential Code* is the latest model building code in the United States, and it appears that most local codes will be based on it in the future. It includes a chapter on ICFs, which really comes from the *Prescriptive Method*. If ICFs aren't covered in the local code yet, your building department might be willing to let you get your specs for the structure of the building from one of these documents. You can order the *Prescriptive Method* from PCA's website (*www.concretehomes.com*), or your ICF supplier might be able to provide copies.

In a growing number of communities the building department is comfortable working with these tables and they know the contractors are familiar with them. In these places they're likely to allow construction for a house or other small building without an engineer's stamp. In communities where ICFs are new, the department may require an engineer to analyze the design and stamp the plans. This is hard to predict, so you should talk to the building officials as soon as you can. It's much more expensive to find out that you have to change the building after it is under construction.

In Canada the situation is similar, but with a twist. ICFs are not yet covered by name in the Canadian model code, the National Building Code of Canada. But the code does give rules for building houses and small buildings out of reinforced concrete. Since ICFs are nothing but reinforced concrete with foam on each face, many engineers and building officials realize that they can use those rules. If they do not, they may be satisfied with the *Prescriptive Method* instead. Also, many ICF companies have "evaluation reports" from the Canadian Construction Materials Centre (CCMC). These are issued by a government organization and include engineering rules for walls built with the company's forms. Hundreds of Canadian building officials and engineers rely on evaluation reports for the engineering specifications for ICF buildings.

In any case, always be sure that the structural parts of the building (concrete and rebar) are designed correctly. If you are not using an engineer you can talk with the ICF company that will supply the forms. Many of them have an engineer on staff who can check your specifications, and almost all of them have knowledgeable staff who have constructed a lot of ICF buildings and know how to work through the tables correctly.

If an engineer will be analyzing the design and stamping plans, you should be able to rely on the engineer's work. The structure of ICFs is nothing more than ordinary reinforced concrete. All licensed engineers are taught reinforced concrete design, so this job should not be difficult. If they need more information, your ICF supplier is a good source. Many ICF companies include engineering tables and information in their manuals.

You can also give them a copy of the *Prescriptive Method for Insulating Concrete Forms in Residential Construction, Second Edition,* or the *International Residential Code.* Either of these will be plenty of backup for most engineers.

If the project does involve an engineer, talk to the engineer early just as you would talk to the architect. Even though engineers have fewer options, there are still things they can do that will make the job easier or harder in the field. Peter, a general contractor in Massachusetts, found this out on a two-story Colonial house he built:

The plans called for two number five rebar over each opening, and they had to go side-by-side. Well, the block we were using was a waffle block, so it gets narrower at points. We had two bars each five-eighths of an inch in diameter right over a cavity that was down to two inches wide at some points. The crew built it the way the plans said—what else could they do? When they poured the concrete, it was almost impossible to get the stuff past those bars over the openings. The guys shook the bars to try to get the concrete from hanging up on them, they cussed and spent a lot of time working on this. After that I called Jerry [the engineer] to ask if there was any way to reduce the bar. He said to go ahead and just use number fours instead, and he signed off on that. It turned out the engineering didn't require number fives at all—he was just playing it safe because he had never done a house in concrete before. I wish I'd gotten him hooked up with the [ICF] company's engineers or somebody else who'd done it before so he didn't overdesign it in the first place.

Preparation

For the first story, the ICFs are set on a footing or a slab. It helps a lot for the footing/slab to be precisely level, because that eliminates the need to do some time-consuming correcting up above. Most contractors say that if the base is level to within $\pm^1/_4$ in they don't have to compensate by adjusting the forms. Short rebar (called **dowels**) usually stick up out of the footing/slab to tie the foundation into the walls structurally (see Fig. 6-4).

The ICF crew puts down chalk lines to mark the position of the walls. Sometimes they also fasten down strips of lumber or other materials along the lines to hold the forms in place. These are called **guides** (see Fig. 6-5).

Figure 6-4 Dowels sticking up out of a footing. (*Courtesy of Reward Wall.*)

Figure 6-5   Guides in position. (*Courtesy of Reward Wall.*)

Before setting forms they also build **bucks** (see Fig. 6-6). A buck is a sort of frame that blocks out the openings for a window or door. Almost everyone agrees that building the bucks ahead makes the job much more efficient because they don't have to interrupt stacking . The crew sets bucks in the wall as they stack it. A buck has fasteners attached that get cast into the concrete to hold the buck firmly in position. Later, the window or door is attached to the buck. Most contractors build bucks out of pressure-treated lumber. But several companies now

Figure 6-6   A buck in position in the wall. (*Courtesy of Arxx Building Products.*)

Figure 6-7 Plastic bucks in position. (*Courtesy of Quad-Lock Building Systems Ltd.*)

make plastic buck material that has its advantages, and many ICF form suppliers carry it (see Fig. 6-7).

On most jobs contractors also build and stand up **corner braces** (see Fig. 6-8). These are vertical posts made of 2× lumber that have a cross section shaped like an "L". One on the outside of each corner of the form wall keeps the corner from

Figure 6-8 A Corner brace in position. (*Courtesy of Reward Wall.*)

pushing out or distorting during the pour. A kicker or two running from the ground to the corner brace holds it plumb.

Setting Forms

There are few hard rules for the order in which the foam forms are put into place to build the wall. Suppliers have different recommended sequences of steps, and most contractors come up with a few wrinkles that make the job more efficient or that suit their people better.

Block walls are built up one row (technically called a **course**) at a time (see Fig. 6-9). Most contractors put down forms for the corners first. They work inward from the corners. On each wall they cut the last block in the middle to fit. Then the corner forms for the second course go on top of the first course, and the contractors stack toward the centers again. This continues to the top of the wall. On each course the ends of the blocks are offset from the blocks below so the vertical seams don't line up. If they did line up, that would create a weak spot in the formwork.

Along the way the workers set the horizontal rebar and bucks in place. When the wall is at the right height, they set and fasten the horizontal rebar on top of the blocks' ties. Or sometimes there are special **rebar saddles** that you set on a block, and the rebar goes on top of them.

When the wall reaches the height of the bottom of a window buck, they cut a notch in the wall to stand the buck in. As they continue stacking, they cut blocks as needed to fit up to the buck. When they stack over the top of a buck they

Figure 6-9 Setting a block in the wall. (*Courtesy of American ConForm.*)

Figure 6-10 The rebar in a lintel over an opening.

usually need to set some horizontal rebar in the blocks just above it. This is to create a structural **lintel** over the opening (see Fig. 6-10). Its purpose is a lot like a header in frame construction—to transfer the loads over the opening to the wall on either side. The horizontal rebar extends a foot or more past either side of the opening.

Because the block scraps can be readily glued back into the wall, experienced contractors generate very little waste on an ICF job. They save the scraps and use them when they see a spot that requires a small piece.

Once the wall reaches eye level, standard procedure is to erect the bracing/scaffolding system (see Fig. 6-11). With the typical system, the crew erects one set (consisting of a vertical post, a diagonal kicker, and scaffold bracket) against the wall about every 6 ft around the perimeter. When the sets are all up, walk planks and rails go in the brackets to create the scaffolding. From the scaffolding the crew can set the higher blocks comfortably.

Once the foam reaches the top of the wall, the crew drops down the vertical rebar. Depending on the local conditions (wind loads, earthquakes, etc.) there could be one vertical bar every 1 to 4 ft along the wall. In addition, one bar usually also goes on each side of an opening. There are various tricks for ensuring that each vertical bar laps the dowel below correctly and doesn't move around. Then the crew lays another layer of horizontal rebar around the perimeter, since nearly all plans call for horizontal bar at the top of a story.

Figure 6-11 Bracing/scaffolding in position. (*Courtesy of ICFA.*)

Inserts

After the form wall is up, it is usually necessary to insert a few things before the pour. If there will be a floor deck at the top of the story, it is almost always necessary to put in connectors for attaching the floor later. When the concrete is placed, it locks in the connectors for a strong structural link. There are easily a dozen different ways to connect a frame floor deck. But probably the most common is still with a prehung **ledger** (see Fig. 6-12). This is 2× lumber or a light-gage steel track that is outfitted with an anchor bolt every foot or two. It is fastened before the pour to the forms along the inside of the wall at floor height. The concrete locks the anchor bolts into position. The ledger gives the framers a secure member to attach their floor to.

A rapidly growing new alternative is to attach the ledger with a **right-angle connector** (see Fig. 6-13). The connector is a steel plate with a right angle bend to it. Workers cut a slot in the foam every 1 to 4 ft and push the long leg of the plate in. That leg sticks into the cavity, so that it will lock into the concrete. The short leg lies against the surface of the foam, exposed to the interior. The floor crew later figures exact floor height and screws the ledger into the exposed short legs of the connectors.

The other major insert is for a totally different purpose. It is a **sleeve**—a piece of plastic pipe let into the wall (see Fig. 6-14). It creates a path through the wall for things like electrical wiring and plumbing. Instead of inserting sleeves before the pour, it is possible to core drill holes after the concrete has cured. But most contractors find that it is less work to plan the places the pass-throughs will go and preinstall the sleeve.

Figure 6-12   Ledger in position on the wall. (*Courtesy of Quad-Lock Building Systems Ltd.*)

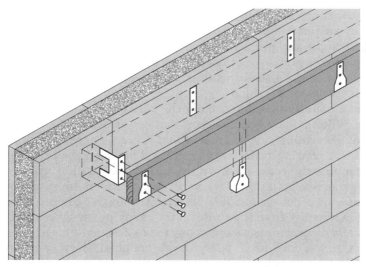

Figure 6-13   Exploded view of attaching a ledger to an ICF wall with a right-angle connector.

Figure 6-14 Sleeves in position. (*Courtesy of Quad-Lock Building Systems Ltd.*)

The Pour

Until they are very experienced, smart contractors leave plenty of extra time before the concrete pour to check everything over. Some ICF companies provide a list of things to double check, and many contractors have made their own lists. Things are easy to change before any concrete is in the wall, and a lot harder afterward.

One key job just before the pour is to plumb the wall. One worker checks the wall for straight and plumb. He does this with a string line along the top and a level placed vertically against the forms. He calls to a second worker who adjusts the kickers of the bracing/scaffolding. They work their way all around the perimeter in this fashion.

Concrete usually is ordered from the ready-mix company the night before the pour. When it shows up the crew had better be ready.

One bit of terminology: engineers and concrete experts talk about **placing** concrete, not pouring it. Don't get confused between the two. With ICFs, both mean dropping the concrete down the wall cavity to fill it up.

Placing concrete in the forms is usually done with some form of **concrete pump**. If the walls are below grade (like basement walls) it's often possible to drop the concrete in, directly from the chute on the ready-mix truck. But even in that situation many contractors now opt to pay the extra money for pump rental. Pumps can reach, so you don't need to have level ground all the way around for the concrete truck. And the pump gives much better control over the flow of the concrete. The sideways rush of material out of a chute can cause problems.

The most popular pump is the **boom pump** (see Fig. 6-15). It has multijointed arms that the driver can adjust to put the end of the line anywhere—up, down,

Figure 6-15   Boom pump in action. (*Courtesy of ICF Accessories.*)

sideways—within a reach of maybe 60 ft. But it puts out the concrete too fast for most ICF forms, so the crews use various fittings on the line to slow it down. The boom pump is also one of the more expensive options.

Second most popular is the **line pump**, also called a **trailer pump**. The line of this pump is a hose that simply lies on the ground. The lead person on the concrete crew holds it over the wall, and a few others on the ground move the heavy line as needed. The line pump runs at a good speed for less money, but involves heavy lifting and requires the extra workers. There are a few other less common options that can be right in particular circumstances.

Concrete safety has become important to contractors and to OSHA. Increasingly, regulations call for anyone on a concrete crew to wear a hard hat, eyewear, and clothing that covers almost every square inch of skin—heavy work boots, long pants, long-sleeve shirt, and gloves. Skin contact with concrete causes irritation (dermatitis). With regular exposure over many years it can cause more serious problems.

The pour operations are organized to place the concrete steadily into the forms without causing high pressures or any abrupt surges. Usually the concrete goes through the window sills first. The sill piece of the bucks has holes to make this possible. The crew then gets on the scaffolding and starts filling from up above. They gradually work around the perimeter of the building, filling the formwork up to about 3 ft high. This partial fill all the way around is called a **lift** (see Fig. 6-16). By the time they get back to the starting point the concrete is already hardening a bit. This reduces the pressure on the forms when they add the weight of the next lift.

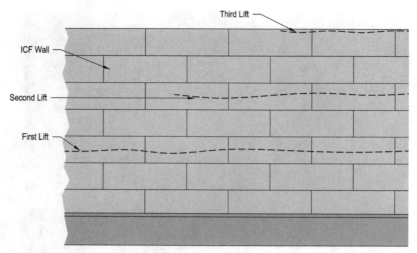

Figure 6-16 Lift Pattern.

As the wall is filled, one to two crew members consolidate the concrete (see Fig. 6-17). The most thorough method of consolidating is with a vibrator. The person handling the vibrator follows along behind the worker holding the hose and filling the forms. He dips the vibrator into the concrete and pulls it out slowly. He repeats this about every 6 to 8 inches, and does it on each lift.

Figure 6-17 Consolidating the concrete in the wall. (*Courtesy of Quad-Lock.*)

Figure 6-18 Troweling the top of the wall. (*Courtesy of Quad-Lock Building Systems Ltd.*)

After each lift two workers check the wall for straight and plumb all around the perimeter and adjust each post as necessary.

If this is a one-story building, when the concrete reaches the top of the forms a couple of workers get on the scaffolding to trowel it level and stick roof connectors into the concrete (see Fig. 6-18).

If this is a lower story of a multistory building, the workers can stop the concrete a couple of inches shy of the top of the forms. The rebar are run a couple of feet long and stick up out of the concrete.

After concrete placement is done, the crew does one final check on the wall for plumb, and adjusts the kickers of the bracing/scaffolding to get it true. They also go through their checklist to make sure everything is now done correctly. A lot of things can still be adjusted while the concrete is wet.

Once the concrete cures, a framing crew comes in to build the floor deck. This is usually done with joists connected to a ledger using joist hangers (see Fig. 6-19).

The upper story goes about the same as the first except the crew works off a floor deck instead of the ground or a slab. Bucks are built first, the forms go up, bracing/scaffolding goes against the walls. The pour is the same, just higher off the ground.

Weather considerations

ICFs are almost ideal formwork for curing concrete because they keep moisture in and they insulate it to keep things at a fairly even temperature. The Portland Cement Association has sponsored tests that indicates that when the outside

Figure 6-19 View from below of floor joists connected to a ledger with joist hangers. (*Courtesy of TF System.*)

temperature is 0°F, concrete placed in the forms stays above freezing for a week and cures to its rated strength.

Jerry, an ICF contractor in Wisconsin, tells:

> We did a basement job for one GC in the dead of winter. We were set to pour, but that day the weather was 30 below, I kid you not. We tried to explain that it was a bad idea to pour in temperatures that low, but he wouldn't hear of it. He was paying us, so what could I say? We did it. The stuff flowed like molasses, but we got it in. I'd never do this again if I have the choice, but the foundation has held up fine.

As Jerry suggests, there's a temperature below which you shouldn't even be putting concrete into ICFs. But having said that, plenty of contractors have stories about placing concrete in subfreezing temperatures without problems. Some have their concrete supplier provide a special mix that accelerates the curing, so it reaches adequate strengths before freezing can occur. Others have the concrete mixed from heated aggregate or water so it comes to the job site warm. Inside the forms it's protected by the insulation. But it's important to cover the top of the formwork with some sort of insulation (pieces of foam, a fiberglass blanket) so that the top surface isn't exposed to the cold.

In hot, dry weather crews often put a layer of plastic sheeting over the surface so that the concrete doesn't dry out. The same trick keeps water out if heavy rain is expected at the end of the pour.

Figure 6-20 Blowout (left) and repair to the blowout (right). (*Courtesy of United Brotherhood of Carpenters.*)

Possible problems

Nearly all errors in ICF construction occur during the pour, or at least show up there. The three that are most talked-about are **blowouts**, **voids**, and **misalignment** of the walls.

Blowouts are actually rare nowadays, and are very easy to fix (see Fig. 6-20). The only reason they are talked about much is that people who don't know ICF construction assume they'll happen a lot. They can't imagine that foam can hold wet concrete. They also assume that blowouts will be a big mess because they don't know how to handle them yet.

During the pour, concrete may burst through a weak point in the forms. It will break the foam and slowly ooze out. This is a blowout. Most experienced crews have, maybe, one blowout every 5 to 10 jobs. They have learned how to build their formwork without weak spots and make the concrete flow so it produces no sudden surges in pressure.

When a blowout does happen, the crew moves the concrete line away from the area of the blowout and continues placing concrete somewhere else in the wall. A couple of workers on the ground pull off the broken piece of foam and clean up the concrete that came out. Then they scoop some concrete out of the wall to get it below the level of the hole, glue the broken foam back in, and cover the spot with blocking or a piece of plywood that is screwed to the tie ends to reinforce the break. That's it. The workers placing the concrete can come back to fill in that spot in the wall at any time.

Voids are air pockets in the wall. Concrete didn't fill a spot and now there is empty space. Voids rarely occur if consolidation is done right. When they do occur you can tell by tapping the forms. The void will sound and feel hollow. If the crew spots this during the pour, they can just go back and consolidate more and add concrete as needed. If the void is discovered after the concrete has cured, it can still be easily fixed. You cut a small hole in the foam at the top of the void and slide concrete into it. When the void is full, you agitate the concrete with a vibrator or stick to make sure it's consolidated and glue the foam cutout back in place to plug things up.

Misalignment means getting walls that aren't straight, plumb, or square. It's actually a more serious issue than the others, and something everyone should keep on watch for. You prevent it by bracing the walls properly and checking and adjusting the bracing before, during, and immediately after the pour. This is all standard procedure, so any crew that is trained and conscientious should be fine. But note that adjusting the walls, like placing the concrete, is a skill that you can't learn entirely out of books. You have to do it. That is another reason to have an experienced person who has done several ICF buildings on-site and calling the shots for your first couple of pours.

Jeff, an experienced ICF contractor in Maine, makes a special point of getting his walls as straight and plumb as possible:

> We run a string line along the top of the wall, one inch out, all the way around. When the pour is done and the concrete is wet, the last thing we do is final alignment. Everybody gets off the scaffolding except for one man. That keeps the weight of the crew from distorting things. Another guy is on the ground working with him. These guys use a level to check the wall for plumb, and the guy on top measures the distance between the wall and the string. Anyplace it's not exactly one inch he calls to the guy below to adjust the kicker. Nothing's perfect, but by the time they're done and the concrete cures we can get walls that are within $\frac{1}{8}$ inch of spec at every point.

Special Wall Features

Curved walls

In the words of Brian, an ICF contractor in Ontario:

> I've done plenty of frame before, and ICFs cost more per unit, but they give me more flexibility, especially in doing curved walls.

ICFs have an advantage on curves because the foam has enough flexibility to bend into an arc. The usual method of making a curve is to cut thin vertical strips out of the inside layer of foam between the ties (see Fig. 6-21). Then you bend the block inward until the cuts close up, and glue it in this position. This creates

Figure 6-21 Cutting a block to curve it (left), and the final curved block (right). (*Courtesy of Arxx Building Products.*)

a curved block, and the wider the strips cut out the tighter the curve. The curved blocks can be stacked to form a curved wall section.

An alternative is to assemble a whole wall section on the ground first, then cut the foam on the inside and bend the whole section. This can be easier than fitting the curved blocks together. The curved wall section gets lifted and set into place all at once (see Fig. 6-22).

Figure 6-22 Curved wall. (*Courtesy of Arxx Building Products.*)

Figure 6-23   Custom-cut corner. (*Courtesy of Quad-Lock Building Systems Ltd.*)

Odd angles

Ninety percent of all building corners are 90°, and these are made with right-angle forms. A few companies offer 45°, 30°, and 60° angle forms for the next most common types of corners. But you can also cut and glue forms individually at whatever angle you want. These custom corners are braced on the outside to hold them steady during the pour (see Fig. 6-23).

T-walls

If one of the interior walls will also be built of ICFs, it joins the exterior walls at a "T" intersection at each of its ends (see Fig. 6-24). The crew can create this by cutting a vertical strip out of the inside layer of foam in the exterior wall forms. The strip's width is equal to the thickness of the concrete wall. Then the forms of the interior wall are butted up to the exterior wall, directly over that cutout, and glued to the exterior wall forms. Lumber braces at the joint and outside the intersection hold everything steady during the pour.

Floor Decks

In multistory buildings there will be a floor deck between each story. Usually this is either wood frame or light-gauge steel frame. However, a growing number of projects now use ICF floor forms. These are made of foam and create a reinforced concrete floor, a little like the way the standard wall forms create a concrete wall. A few other types of concrete floor systems are used occasionally, too. Concrete floors are especially popular for large buildings and commercial construction.

Figure 6-24 T intersection in the form wall. (*Courtesy of Quad-Lock Building Systems Ltd.*)

Frame

Both wood and steel frame decks are usually attached to an ICF wall with a ledger. If the ledger is in position and cast into the wall during the pour, the framers simply connect wood joists to the ledger with joist hangers.

Steel ledgers are usually steel track. The steel joists slide into the track and are screwed through the top and bottom flanges of the track, and may also be attached with a steel clip angle.

If the floor is attached to the wall with right-angle ledger connectors, the framers screw the ledger to the connectors sometime after the pour. Then they attach the joists to the ledger.

ICF

New and growing in popularity are **ICF floor decks**. These are made with long forms, called **sections**. The sections are foam with light-gauge steel joists embedded inside. The steel stiffens the foam so it can span a few feet and still hold the weight of wet concrete. The foam is shaped with a continuous trench, called a **beam pocket**, running the length of each section.

To create the floor the crew sets the sections on top of the cured ICF wall. A line of shoring goes underneath about every 6 ft to support the forms. The crew puts rebar into the beam pockets and a layer of welded wire mesh or additional rebar up above the top surface of the foam. A flatwork crew casts concrete over the forms and finishes it to level (see Fig. 6-25). After it cures the shoring can come out and work can begin on the next story of walls.

Figure 6-25   Putting ICF floor forms in position. (*Courtesy of INSUL-DECK LLC.*)

Other concrete

There are many methods other than ICF floor forms to build a concrete floor. All have their advantages and are used at least occasionally. They include the following:

- *Composite bar joist system* (see Fig. 6-26). The crew sets a special heavy steel bar joist on the ICF walls, puts plywood between the joists, sets welded wire mesh on top, and pours the concrete. The plywood drops out later. There's a lot of freedom to run cable and ductwork between or through the joists.

Figure 6-26   Formwork for a composite bar joist system, seen from below. (*Courtesy of Canam Steel Corporation.*)

Figure 6-27 Setting precast hollowcore plank in position. (*Courtesy of Integraspec.*)

- *Traditional form deck.* This is built like an ICF floor deck, but with plywood as the form instead of foam sections. The plywood is removed after the concrete cures.

- *Precast hollowcore plank* (see Fig. 6-27). This is a thick section of concrete with hollows inside that is made at a factory. It is delivered to the site and hoisted into position with a crane. The crane sets it on the lower-story walls, and the crew ties it into the walls with rebar.

Some other, less-used types of concrete floors are also available, and some promising new ones are under development.

Roofs

Most roofs on ICF buildings are also wood or steel frame. Recently in some projects the roof has been constructed with the same new ICF deck formwork used on some floors.

Frame

Most frame roofs built are conventional, either preengineered trusses or stick-built rafters and joists. In high-wind areas, these roof members connect to the top of the ICF walls with a steel strap. The straps are commonly called **hurricane straps** (see Fig. 6-28). The crew embeds the base of the straps in the concrete when they place it in the walls. This leaves a free end sticking up. After the cure the framers set the roof members and fasten one or two straps (depending on the strength needed) to each one. According to Bob, a contractor in Iowa:

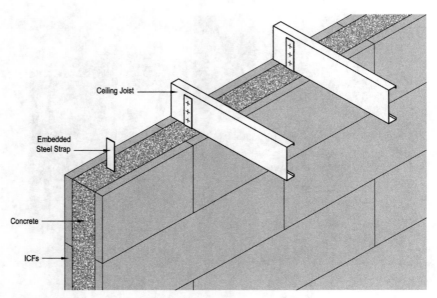

Ceiling Joist

Embedded
Steel Strap

Concrete

ICFs

Figure 6-28 Hurricane strap in position.

There was an ICF house hit by a tornado in Iowa. The builder tied the roof to the concrete walls with hurricane straps. The result was very little damage (it lost some shingles and some vinyl siding), even though neighboring houses were totally wiped out.

In areas with lower winds, most roofs are connected by way of a **top plate** instead (see Fig. 6-29). Using a top plate adds some flexibility because you don't need to know the location of the trusses/rafters when you place the concrete. But it is usually a slightly weaker connection. The ICF crew embeds either sill straps (similar to hurricane straps) or anchor bolts in the top of the wall. They use these to fasten down a plate of 2× lumber or light-gauge steel. The trusses or rafters then get attached to the plate the same way they attach to the top of a frame wall.

ICF and Other Concrete

It's possible to create a concrete roof with the same systems used for a concrete floor. If the roof is flat, the forms are set and the concrete is cast the same way they are for a floor. If the roof has a pitch, the job is more complex. With ICF deck forms the bracing will have to be set at increasing heights to support a slope (see Fig. 6-30). The sections need to be secured so they do not slide off. They also need to be specially cut to create a deep beam pocket along the ridge of the roof. The pour requires a fairly stiff concrete and an experienced crew. Roofs with pitches of 4-in-12 or less have been done often. Greater pitches are much more difficult and very rare.

In theory a pitched roof could be constructed with any of the other concrete roof systems, too. For each system this will probably require more planning and extra work for things like ridge beams and miter cuts.

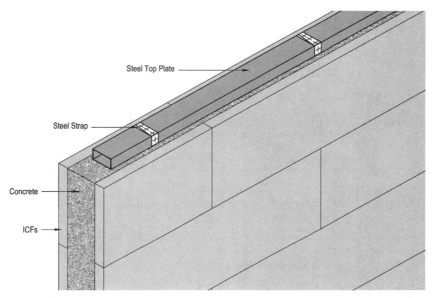

Steel Top Plate

Steel Strap

Concrete

ICFs

Figure 6-29 Top plate in position.

A third option that gives many of the advantages of a pitched concrete roof without the complexity is a **safe ceiling** (see Fig. 6-31). This is really a concrete floor deck put on top of the house. A pitched frame roof goes on top of that. The floor deck goes up in the usual way, except that connectors are embedded around the perimeter. Then frame roof trusses/rafters get fastened to the connectors the

Figure 6-30 A pitched ICF roof ready for the pour. (*Courtesy of Lite-Form International.*)

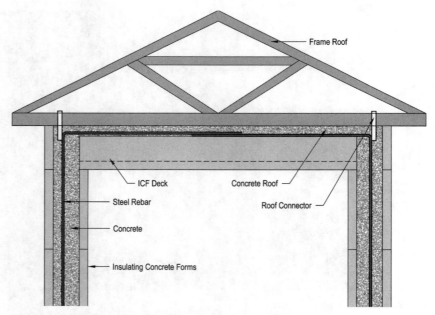

Figure 6-31   Diagram of a safe ceiling.

same way they get fastened directly to ICF walls. The result is a main house "box" that has the energy efficiency and wind resistance of a total ICF shell, with a full-pitch roof and attic space on top.

Safe Room

A **safe room** is a single, wind-resistant room that gives occupants of a conventional building some place to ride out a hurricane or tornado (see Fig. 6-32). A walk-in closet is a popular choice for the safe room. It is large enough to hold a few people, has no window and only one door. Although it is possible to build safe rooms out of different high-strength materials, ICFs are great for the job.

Safe rooms are nearly always constructed on a slab foundation. Steel rebar dowels in the slab extend into the safe room walls to make a high-strength tie-down for the safe room. The ceiling is also concrete, formed with ICF deck systems. The wall rebar extends up into the deck to lock everything together. A heavy-duty steel door completes the package.

The Federal Emergency Management Agency (FEMA) publishes a guide to the construction of safe rooms that includes plans for building them out of ICFs or other materials. It's called "Taking Shelter from the Storm: Building a Safe Room Inside Your House." Go to the FEMA website (*www.fema.gov*) to download it for free or get the information for ordering a hard copy.

Figure 6-32   Typical safe room ready for concrete.

Radiant Heating

A growing number of ICF buildings call for concrete floors with in-floor radiant heating. In-floor radiant heating is a fairly expensive upgrade in a frame building because it requires that the floor be topped with gypsum or concrete. There is a cost to that, and in addition to this the framing must be beefed up to carry the extra weight. But in a building the owner is outfitting with ICF walls and a concrete floor anyway, the heating tubing can be embedded in the floor's concrete with little or no extra measures (see Fig. 6-33).

To install radiant heating, workers lay down plastic tubing across the floor forms before any concrete is cast. If the floor gets a top layer of welded wire mesh or rebar, that goes on next. Then the pour and slab finishing proceed as usual. With the tubing cast in place the heating installers can hook the tubing up to the heating system. Warm water pumps through the tubes to heat the slab and warm the building.

Electrical and Plumbing

Before the wallboard goes on, the electrician puts cable and boxes in the foam face of the wall (see Fig. 6-34). Plumbing in exterior walls isn't common anymore, and in many areas it's frowned upon. But if desired, it can and does go into ICF walls about the same way as the electrical does.

The electrician cuts chases and rectangles out of the foam to hold the cable and boxes. Many different tools can do this efficiently. But one favorite and the most-recommended tool is the hot knife. Workers bend the blade in a loop shape and it makes clean, consistent chases in one pass. The electrician can cut a rectangle with a few strokes of a straight blade.

Figure 6-33 Radiant floor tubing in place on an ICF form deck. (*Courtesy of Quad-Lock Building Systems Ltd.*)

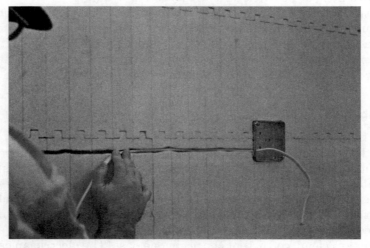

Figure 6-34 Installing electrical cable. (*Courtesy of Portland Cement Association.*)

The electrician lays the cable in the chase—there is no need to thread it through holes. He may spot glue it with foam adhesive every foot or two to hold it securely at the back of the chase. Boxes should fit precisely in the rectangular cutouts. They may be glued to the foam, but some inspectors and some contractors prefer to have them screwed into a tie or fastened to the concrete in back. Pulling the cables through the boxes and the remaining electrical work go the same way they do in a frame building.

Plumbing pipes go in much the same way. A large-diameter pipe like a vent or drain may not fit in the inside foam layer. In that case room for the pipe may be gained by boxing around it with some studs.

Interior Finish

Ninety-nine times out of a hundred, ICFs are finished on the interior with conventional gypsum wallboard, or one of its close substitutes (plaster board for a full plaster job or fiberglass-cement board for water resistance). The crew attaches these to the ICF ties with wallboard screws just the way they attach them to the studs of a frame building (see Fig. 6-35). Occasionally you'll find a crew that uses glue in addition to screws or instead of them, but the standard screw works fine.

Exterior Finish

All the standard finishes attached to the outside of frame can go on ICFs, and they go on about the same way.

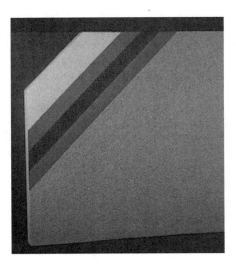

Figure 6-35 Wallboard installation in progress. (*Courtesy of Portland Cement Association.*)

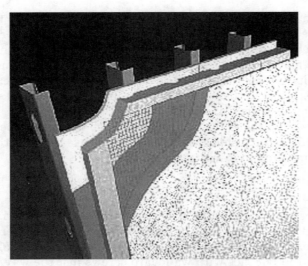

Figure 6-36 The layers of an Exterior Insulation and Finish System. (*Courtesy of Dryvit Systems Inc.*)

Exterior Insulation and Finish Systems (EIFS) are a popular exterior finish over ICFs. Although they are often compared to stucco, EIFS are actually an acrylic–based material with high impact resistance that can have a wide range of colors, including very bright ones. One of the reasons for their popularity on ICFs are that they may go onto an ICF wall even more easily than onto frame. Conventional EIFS includes a layer of EPS board that gets adhesively or mechanically attached to the surface of the wall. This is followed by a trowel- or spray-applied acrylic base coat and a layer of fiberglass reinforcing mesh that is embedded in the base coat. Finally a layer of integrally-colored, acrylic, textured finish is trowel or spray-applied. An ICF wall may not need the layer of EPS if it is true and plumb. The other materials go on as usual.

Traditional stucco (sometimes called **hard-coat stucco**) is also popular and easy to install. It has little or no plastic in it—usually just cement and fine sand. Over frame, the installation starts with a layer of a water-resistant paper or film. Crews then fasten a wire mesh to the studs and trowel coats of the stucco over the mesh. Again, ICFs don't need the paper layer because the foam is a perfectly good substrate already. The mesh fastens to the ties of the forms, and the other things go on about the same as with frame.

Sidings, like vinyl, clapboard, and hardboard simply nail or screw to the ICF tie ends. The only caution is that they must break on a tie. When using nails, good contractors stick with ring shank or galvanized nails because those hold a better grip in the plastic ties. With steel ties, the general rule is to stick with screws.

Masonry veneers, like brick, stone, and architectural block, go on the same way they do on frame. Steel brick ties are fastened to the ICF tie ends and the masonry crew stacks the veneer, embedding the brick ties in the mortar joints as they go. The masonry sits on the slab or a brick ledge, just as in frame construction.

Choosing Suppliers

Overview

One of the biggest wastes of time for the new ICF contractor is trying to figure out which is the best ICF system to use. Some of these guys can talk forever about which ICF has foam that's one-quarter inch thicker, whether it's better to have a tooth or a tongue-and-groove on the edge, and why polypropylene ties are better than polystyrene ties or vice-versa.

It's not that these things are totally unimportant. There are differences between systems that can make installation a little faster and cheaper, and certain crews like some features because they fit better with the way the crew likes to work. What's wrong with spending a lot of time on these questions is that they are *secondary*.

It is much more important *who* you get your forms from. You'll need a lot of technical support, fill-in orders, and marketing materials down the road. The ICF manufacturers provide this stuff in abundance—or at least, the good ones do. It only takes a little time to make sure the products you're considering are OK. There are hardly any really bad ones anymore, anyway. Spend more time figuring out who can service you the best.

Also spend a little time considering your other supplies. Your forms are the most important material, but there are a few other vendors you're going to need to find too. Nowadays there are some equipment, tools, connectors, and supplies designed just for ICFs. These can come in very handy and you need to know where to get them.

Rule Number One—Talk to Contractors

If you've been following our advice, you have already found some local ICF contractors. If there aren't any near you, find the closest ones you can. And make sure you talk to them *about products and suppliers*. Whose forms do they buy? Why? What equipment and accessories do they use? Why? Where do they get them?

We told you to talk to these contractors to find out about the local market and getting into ICFs. Most are happy to talk to you about their suppliers, too. Buy them dinner if you have to. It will be well worth it.

It's also helpful to check out the Internet chat rooms where contractors talk ICFs, like the one on *www.icfweb.com*. The topics can be a bit random, but people are mostly telling you what they really believe.

This information is very, very useful, but *be careful*. You have to take contractors' opinions with a grain of salt. Some people have a bad experience with a product or a company, even though another 99 are happy with it. Some have different types of operations from yours, so what makes them like a supplier would not always be useful to you. And some people just plain exaggerate things.

And what about the companies you never hear about? Maybe you never happen to talk to anyone who buys from a perfectly good company—that doesn't mean you shouldn't consider it. So you can't just believe the person who yells loudest or who you talked to last. Combine the information you get from other contractors with the information you get from the companies themselves and your own observations.

Rule Number Two—Leave No Stone Unturned

It's easy to find out which ICF companies operate in your area. It's also easy to get information from them. They want your business, so they'll come to you if you ask. Find them all and get their information.

To locate the major companies, a great source is *www.forms.org*, the web site of the Insulating Concrete Form Association. Just about every ICF company worth talking to is a member, and they're all listed on the site. Go to *Member Search*. The ICF companies are called *Primary Members*. And don't just look at the ones listed in your area of the country. They're classified by where their headquarters are, but most of them cover a wide area. Probably half sell all across the United States and Canada.

So contact them all. It's easy. There is an e-mail address for each one. Click on it and send a message explaining that you are considering building regularly with ICFs. Ask for a local contact or distributor. They can send literature and have their rep get in touch with you.

Product

Since you'll probably spend a lot of time comparing the specifications of the different products no matter what we say, let's get it out of your system right now.

The fact is that the final ICF wall has about the same properties regardless of which system it's built with. They're all approximately 2 in of foam on each side with reinforced concrete in the middle. But the wall gets assembled a bit differently with different products. Some contractors feel that certain form designs or certain product features help them build the wall a little faster or more reliably.

The problem, of course, is that these contractors disagree. They don't consider one particular brand *the best*. Which form you use simply depends a lot on personal preference and your own style of operation.

So go ahead and compare all the products feature by feature. Make a list of the things you like and dislike about each one. Get other people's opinions about the advantages and disadvantages of each feature. Just don't let it rule your thinking. There's little advantage to using the "best" product if you can't get it delivered locally, or no one is there to help you when the building department is asking questions.

Support

If there is one overriding thing to look for in your ICF supplier it is *support*. When you look for your supplier, tattoo this word on the back of each hand to remind yourself.

Randy, a contractor in Georgia who recently converted to ICFs, said:

> I was building my own house in the mountains of Northern Georgia. We got great support from my local ICF distributor. We received a total of 15 hours of on-site training and supervision. I think that was very helpful and key for our first job.

You want the same kind of help Randy got, especially on your first job or two. ICFs work well, but even so, you will have dozens of questions pop up when you're building. Do we mount doors toward the inside or outside of the wall? Do we build or brace the inside corners any differently from the outside corners? What if we don't have a vibrator to consolidate the concrete? The rebar as shown on the plans would have to go right through this new opening the customer asked for—what do I do about that? You really want a company that returns your calls and knows the answers.

On your first few jobs you may also have a lot of different people asking you more questions—the owner, the building department, your concrete supplier, various subs, maybe an architect, maybe an engineer. It helps greatly to have someone who's been through this before to call these people and tell them exactly what they need to know. It has to be someone who knows his stuff and can speak their language, or they'll get nervous.

A lot of ICF companies are armed to the teeth. They have a distributor in your area who knows how things work there. The distributor can even come to the job site at critical points. They have plenty of literature with facts and graphs and engineering tables. They have evaluation or research reports from the major model code bodies—International, BOCA, Southern, or ICBO. They have marketing literature that shows buyers exactly what they're getting and how valuable it is to them. They have engineers on staff who can talk a designer or building official through a problem or issue.

You want all this stuff. Ask about all these things at each company you're considering. Ask for copies of their literature. You'll probably find several companies that are strong in all these areas, and that's fine. You can weigh in the other factors and pick the one that looks best, all things considered.

After ten projects you may find yourself using this support less. If that's the case, you can certainly consider switching to a different company that provides less of it. But don't think that way for your first few projects. You'll need it then.

Geographic Availability

Shipping costs can add up, but the truth is that major manufacturers of ICFs all now have several plants around North America to serve their customers. So it's rare that you have to worry about whether an ICF company has a plant close enough or not.

What *can* be important is whether they have anyone who stocks the product locally. So long as you order a couple of weeks in advance, you can usually get an order of forms from the most far-flung producer at a shipping price of only a dime or two per square foot. But when you find out that you're a couple of blocks short at the top of a wall, you start to appreciate the value of quick delivery.

Consider the rest of the story from Randy, the contractor from Georgia:

> Everything was going fine until the end when we came up 40 blocks short and had to wait for 2 weeks for delivery. The distributor did not have immediately enough in stock and we either had to pay for the shipping from the manufacturing plant in another state or look for "leftover blocks" at other local distributors. We ended up choosing the second option and had to sort through a lot of blocks to pick up the better ones. Even though we got a very good deal on those it was kind of a hassle.

So check which local distributors inventory product. For those that don't, can they get product from someone else nearby who keeps an inventory? What sort of backup do they have in case the local stock is out?

Training

This is so important that this book has a whole chapter on it. Whatever product you use, make certain that you get well trained before, or during, your first project. Whoever you buy from, you want to make sure there is training available for you on their system.

Some companies have their own training programs. Their own instructors run the course, and they rotate it so programs show up in each part of North America often. A great solution to the training problem is to find a company that has a comprehensive course at a time, place, and price you can handle.

But don't rule out a company that doesn't offer a formal course. Many provide a trained person to work with you at length on your first project. That is extremely valuable. In fact, if you had to choose between a company with a course but no on-site support and a company with on-site support but no course, you should probably choose the on-site support. And, as you'll see in the next chapter, there are some good training programs that are offered by independent organizations that cover a lot of different ICF systems. So you can often take a course even if your form supplier doesn't offer one.

Just make sure before you commit to a supplier that there is *some* good way to get fully trained on its product.

Price

The price of the forms can be very important, but not always for the reason you would think. If you can buy forms that are 50 cents a square foot cheaper, that could save one or two thousand dollars on a typical house. That's obviously an important consideration. But if that less expensive form comes with a fraction of the technical support or training, it is probably no bargain. You will pay back that thousand or two and more in extra labor, waste, lost time trying to reassure the building inspector and owner and designer, and rework on things that should have been done differently in the first place.

The same goes for a bargain on your first job. Some distributors may offer you a discounted price to get you to try their product, and that is fine. There's nothing wrong with getting a good first deal from a company that wants your business. Just make sure it's with a company you want to be working with—one that has the training and support and availability.

Remember that you can always switch brands down the line. There are a few things to learn with a different system, but they're pretty minor. The basics are all the same. If you start out with a full-service firm and you realize after a few jobs that you're not using all of its services, you can start buying reduced-price product from a "discount" company. And that's what these companies are for— lower-priced product for someone who doesn't need the extras.

The truth is that many contractors discover they need the service much longer than they imagined. They find that they are constantly doing new things— more elaborate houses, unusual details, a wide range of commercial projects. They have new questions and they are constantly dealing with new people who need to be reassured. The support from the ICF supplier may be important to you for a long time. But rest assured that you can switch later if it makes sense.

For your information, as of 2003 a pretty typical price for forms was about $3.00 per square foot of wall area, plus or minus a bit. Prices close to $2.00 crop up occasionally, but most of them do not include shipping (which adds a dime or two) and come from companies that provide lower levels of support.

Lower prices may also be found on products that have fewer features and require some extra labor in the field. This can be fine, but you should make sure you know all the extra steps you'll have to carry out, and have some idea what extra cost of these extra steps will be.

Quality

Fortunately, you don't have to worry about product quality much nowadays. Manufacturers have now produced billions of forms, and their control over quality is extremely good. But go ahead and check on product quality anyway. It is important. Ten years ago some forms had inconsistent dimensions. Some forms in a batch might be, say, half an inch long. This doesn't sound too bad, but it makes stacking them into a wall slow and frustrating. And some forms had occasional weak spots that might give out during the pour. According to Kurt, an ICF contractor in Utah:

> The difference between the forms a long time ago and today is like night and day. I tried ICFs for the first time 11 years ago but had a horrible experience. The forms were poorly manufactured and didn't hold the concrete. I tried them again two years ago because a customer requested them. Personally I hated them and I did not have high expectations because of my past bad experience. But the customer wanted them for their structural properties and energy efficiency. But what a pleasant surprise they were. They held up like a champ and everything went up easy and came out straight and plumb.

Now Kurt is a regular ICF builder. For your own peace of mind, ask to pick through a pile of forms if you can and measure them to make sure they're consistent. Ask contractors who have used a system if they've had any dimensional or blowout problems. Nine times out of ten everything will be fine, but check anyway. And by the way, *don't* listen to contractors who haven't used the product recently. They go on hearsay or what happened several years ago.

Related Products

Most ICF companies sell a variety of specialty forms and accessories. These can be useful, and you might prefer a company that has a broad product range. But before you turn your back on a company that doesn't have all these extra products, you need to know that you can get many of these things from other sources, too. In the end it can be more important to know what your local distributor *carries* than what your national forms supplier *makes*.

Useful related products that some ICF companies supply are as follows:

- 90° corner forms
- Brick ledge forms
- Odd-angle corner forms
- Bracing/scaffolding system
- Plastic buck material
- ICF deck forms

A 90° corner form is almost essential (see Fig. 7-1). Every project has several corners, and it can be very time consuming to make your own by miter cutting straight forms. So it's well worth it to buy the corner forms. You might steer clear of any company that doesn't have them. And you can't buy corner forms from other companies, because most forms from one company don't fit with the forms from another. But as you might expect, almost every ICF company has them.

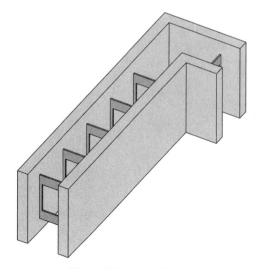

Figure 7-1 Typical 90° corner form.

There are also special forms for easily making brick ledges and certain non-90 corners (see Fig. 7-2). Whether you need a company that provides these depends on how often you build these kinds of wall features. If you put a brick ledge near the base of every building you construct, having that prefabricated brick ledge form can save a lot of time, money, and scrap, and provide a better quality ledge. Making your own ledge by cutting and pasting standard forms is bearable if you only have to do it occasionally. But it's inefficient if you have to do it regularly.

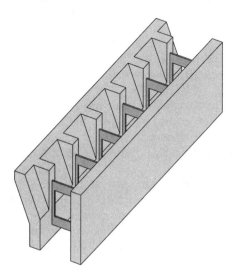

Figure 7-2 Typical brick ledge form.

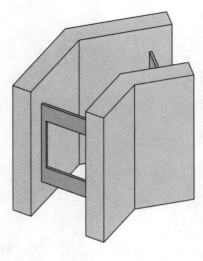

Figure 7-3 Typical 45° corner form.

Odd-angle forms are just like regular corner forms, but for corners that are something other than 90°. They come in 30°, 45°, or 60° versions. If you do, say, a lot of bays, the odd-angle forms are very useful. But if you do them rarely, this is a very minor consideration in deciding what company to use. For the once-in-a-blue-moon that you have to produce a 45° corner you can just miter cut straight forms and glue and brace them (see Fig. 7-3).

A good bracing/scaffolding system is a must (see Fig. 7-4). It aligns the form-work, holding it plumb and square until the concrete cures, and it provides a secure platform for work on the upper parts of the wall. Some ICF companies supply their own designs. But there are a lot of places to get them other than your form supplier. Several independent manufacturers make very good bracing/scaffolding systems you can buy. And almost every bracing/scaffolding system—whether it's sold by an ICF forms company or not—will work on every major brand of ICF. But wherever you get it, check to be sure it meets OSHA regulations.

Plastic bucks are quite popular instead of building bucks out of wood. And some ICF companies sell their own brand of plastic buck material. However, there are independent suppliers of buck material that make and sell it, too. It might be nice to have plastic buck material made by the maker of your forms, but it's not essential.

A popular new product is the ICF deck form for making reinforced concrete floors and roofs (see Fig. 7-5). These work much like ICF wall forms. They're made of foam with light-gauge steel joists embedded inside. The contractor puts them in place for a floor or roof deck, installs the steel rebar, and casts concrete on top. The foam and joists stay in place. Several ICF companies sell these forms. However, they are also available from independent producers, and any type of deck form can work with almost any company's wall forms.

Figure 7-4 Bracing/scaffolding system in place. (*Courtesy of ICFA.*)

The bottom line is that you need a company that supplies right-angle corner forms. There are also definite advantages to a company that supplies brick ledge and odd-angle forms, especially if you do a lot of these things. But the other products are available from many sources, and nearly all of them work with anyone's wall forms.

Figure 7-5 ICF deck forms in place. (*Courtesy of Lite-Form International.*)

It is often more sensible to ask whose local *distributor* has a broader line of products and equipment. Say that one ICF company (Call it "Company A") that has its own bracing/scaffolding, plastic bucking, and deck forms, has a distributor in your area. And say that another ICF company ("Company B") that has none of these things also has a distributor in your area. Company A's distributor may well have a broader product line because he carries all the different things that Company A sells. But Company B's distributor may have all these things, too, because he gets the bracing/scaffolding, bucking, and deck forms from independent suppliers and sells them all to you. Whose supplies are better is something you'll have to decide by examining their products one by one.

And remember, too, that several independent companies now sell a broad range of ICF-related products, tools, and accessories. So you can certainly get your wall forms from your favorite ICF company and some or all of the related items from a separate supplier.

All of this advice holds for a variety of other useful items, as well—low-expanding foam adhesive, foam applicator guns, rebar bender-cutters, cable ties, and so on. Many forms distributors carry these popular supplies, and it can be handy to get these from the same source that sells you your wall forms. But the independent suppliers sell all of these, too, and may even have a broader selection.

The Decision

Before you decide which product you want to use for your first project, take a deep breath and make a list. It doesn't take long. List the half-dozen things you think are most important to you in the company you buy forms from and the product it supplies. We've told you what we think most of these items should be, but you have to decide for yourself what's most important in your case. Then write down each company you're considering, and rate them on each of the things that are important to you.

You don't have to use a fancy scoring system or add things up to get a total. The important thing is that you have to *think* and you have to *be honest with yourself*. What factor do you consider most important? Why? Did other contractors confirm that it's really important? How do you rate Company A on each item? Do you have any evidence, or did you just rate A high because you like the distributor's truck?

This way you can base your decision on facts. And before you stress about this too much, remember that you can change brands if something goes wrong with your first choice. It's not really ideal. It would be cheapest and easiest to hit on the best one for you the first time, but it's not a major effort to learn a second system once you've learned one.

The truth is, the ICF suppliers have become a very careful, well-financed group of companies that have refined their products and their product support several times. You want to develop a good relationship with a good company, but that's not hard to do, and a good company is not hard to find.

8

Training

Overview

No matter what else you might do, at least make sure that you have experienced help on your first job, and you get at least basic instruction yourself before you get there. According to Larry, a new ICF contractor in Georgia:

> I have been a framing contractor for many years and went through the standard ICF classroom training. When I got to the job site though I realized that building with ICFs is not just stacking forms. Basically we were finished with it in 3 days, but we had to learn a lot of new things as we went. Especially when it comes to bracing the walls and putting up shoring for the ICF deck. The ICF distributor that we were working with gave us great support. He came on the job site several times to show us the basic steps first and he came to make sure everything was correct before we poured the concrete. Later we did the second story in half the time. Learning on the site can't be substituted with any classroom courses. You can learn the general information there but until you touch the forms and see how an experienced builder does things you can't fool yourself that you know how to build with ICFs.

It is possible to get more training than you need, but more new ICF contractors make the opposite mistake. They go into their first project with a little knowledge of the product and a lot of confidence that they can figure anything out. They do figure the ICFs out, but with more errors, delays, waste, labor, and rework than they needed to have.

Getting adequately trained is not terribly difficult or expensive. But it takes some effort. Ideally, the person who runs the ICF crew gets trained, and ideally, this person does three things:

1. Goes through a formal training course.
2. Reads the manual in advance.
3. Has an experienced person help out at the site on the first project.

It is great if some of the rest of the crew can get trained, too. That way the lead person has to do less explaining. In a large construction company, it's great if some of the upper management can take a course. Then they will understand what the field crew is doing and will be able to work with them more effectively. But the one critical person is the leader of the crew. No other person can make up for a foreman who has to feel his way with the product. As you read the rest of this chapter, remember that when we say "you" should do this and that, we mean *at least* the supervisor of the ICF crew. More is nice, the supervisor is critical.

It may be difficult for the foreman to perform all three of the ideal steps of training. There may not be a formal course running before it is time to start the first project, and manuals can be hard to read after a long day in the field. However, there are some ways you can cover for these activities—other things you can do to make up for not doing the recommended things. To repeat ourselves, If you can't do everything you should *at least get someone experienced at the job site*, and try your best to *find some way to get basic instruction before you get there yourself.*

Training Courses

There are a lot of things to be said for going through a formal ICF training course. The good courses compress all the experience of hundreds of ICF contractors and then present you with the things that have worked best for them and the things you should really avoid. You don't have to go through all that trial and error. The courses also give you the *reasons* for why you should do things certain ways. This can be very useful. Most of the time your job will not be exactly like the typical job, or exactly like anyone else's job. But if you know why things are supposed to be done the way they are, you can reason out what you should do in your specific situation. And there are certain numbers and names and facts and formulas you will need. It can be very efficient just to go through these in one organized sitting. And the time and trouble you save on your first project alone will probably pay you back for the time and money invested in the class.

A lot of the ICF companies now offer courses in the correct installation of their systems. They only talk about their own system, of course. But that allows them to be very detailed. The typical manufacturer's course is 1 to 2 days. The company's trainers usually travel, so they can offer the course in, say, the Southwest in March, give it in various other parts of the country after that, and come back around to the Southwest to give it again a couple of months later. The cost is usually $100 to $300, plus whatever you pay to travel to the class and stay nearby. When you're looking into ICF companies, ask which ones have a course, when it's offered, and what the terms are.

If you're a union carpenter, you're eligible to take the new United Brotherhood of Carpenters course. It is designed to qualify union carpenters to work on any of the major types of ICF systems. Availability and delivery may be a bit different from local to local, but it's an excellent program. Check with your local to see if you can take it.

Additional courses are available in Canada. The Cement Association of Canada has developed a complete set of materials for training ICF contractors that have been picked up by Canadian trade unions, community colleges, and other organizations. Call the CAC directly (613-236-9471) to find out who in your area is offering a course with these materials.

Manuals

Most ICF manufacturers supply a manual for their products. Most of these are very thorough. Always get the manual. It's free and it can only help.

If you have taken the manufacturer's course, read the manual after the course and before you build. It will be mostly review, and will go quickly. Reading it will remind you of things you've forgotten and fill in some information. This is a great way to prepare for your project because it gets you to think through a lot of details.

If you have taken a general course in ICF construction, reading the manual will also give you all the specifics about your particular system.

If you are going to try to build without taking a course, devour the manual. It will be slower going because everything will be new. But do your best to read and understand every detail. This will give you important knowledge to fall back on when you are puzzling things out in the field. And keep the manual around as long as you use the product. There are always a few things you forget or need to look up later. In fact, for your first few jobs, make it a habit to bring the manual to the job site.

Self-Training Materials

If you can't attend a formal training course, some materials for learning on your own can help. Relying on them is second best, but it's better than going to the job site cold.

The Portland Cement Association sells a set of tapes that lays out the steps of ICF construction and shows you actual crews carrying them out. You can find it on their web site *www.concretehomes.com* under product code VC-500.

There are some printed materials you can read, although they won't have a lot more information than most manufacturers' manuals.

Insulating Concrete Forms for Residential Design and Construction (published by McGraw-Hill, 1997) and *Insulating Concrete Forms Construction Manual* (McGraw-Hill, 1995) are two comprehensive overview books on ICFs. They are available from any major bookseller or on the *www.concretehomes.com* site.

The ICF Answer Book for Builders/DIYs (BuildCentral, 2001) has a lot of useful information on specific construction tasks that aren't always covered in the manuals. It is a collection of answers that experienced ICF contractors gave to questions from new contractors and homeowners. It's available on *www.icfweb.com*.

The Internet also has training information. On *www.icfweb.com/learningcenter/* there is an overview of the ICF systems, an online construction manual, a set of frequently asked questions, and a discussion board that you can use to get some of your questions answered by other industry participants. Most of the manufacturers' web sites also feature step-by-step installation procedures, but those are not always as thorough as the ones in their printed manuals.

Site Assistance

You are not fully trained until you actually build something. Some skills you need to build an ICF wall correctly and efficiently can only be acquired with experience. The greatest of these skills is placing the concrete. You can read instructions all you want, but you will never be able to do it well until you've held the hose for a few hours and filled a wall or three. Basic form assembly also goes more slowly and with more rework until you have done it a few times.

It is *very* important that someone experienced be on hand for at least your very first job. Consider it the final phase of your training. If you simply read or hear how to do the job, you will spend a lot more time scratching your head and redoing things. During form assembly you may get near the top of the wall and suddenly realize that you have to disassemble things and redo them. During the pour things can get sadly out of alignment if you have not prepared bracing correctly and placed the concrete at the correct pace. Two days later you may realize that you dimensioned the openings wrong or embedded some connectors a couple of inches away from where they should have gone. Doing all these things correctly is second nature to somebody who has done a few ICF buildings from beginning to end.

So always get an experienced person at the job site when you start stacking the forms, and again for the pour. You need someone at the beginning to make sure you are stacking correctly. Someone for a full day is ideal, but half a day should be enough if you're organized. That person will pick up anything you're doing that could cause trouble later and he can give pointers for doing the rest of the stacking correctly.

The pour is so heavily dependent on judgment and "feel" that you should have experienced help through the whole thing, without question. An assistant should be there early, at least a couple of hours before concrete is scheduled to arrive. That way there is time to clean up any errors in the forms or cancel the ready-mix if things are really flawed. You can get advice and help in placing the concrete, and learn how to direct the flow and pace the pour.

With this kind of help you should be safe to do the next building on your own. Only if you try something very different from any ICF job you've done before might you need to go back and get more on-site help again.

Nine times out of ten you can get this on-site help from the company that sold you the forms. In fact, that should be one of the factors you used in picking your forms supplier (that they provide on-site assistance). If it's clear that you intend to do ICF construction regularly, most distributors will send someone to the job site to work right alongside you. They will be happy to have you as a customer. They will want you to be satisfied, and they will want you to be well trained so your work reflects well on the product.

When you're deciding whom to buy your forms from, ask about their policy on help at the job site. When you talk to other customers of theirs, ask how good the company was about getting someone to the job site when asked. Try very hard to favor those companies that provided the best site assistance. Then make it clear to the local distributor from day one that you will be asking for experienced help on your first project. Keep in touch about the schedule as the day approaches. Then make sure your help shows up, and try not to go too far if it doesn't. Spend more time calling to nag your helper into coming and less on actually building anything before then.

If you can't get a site assistant from the manufacturer for any reason, talk to other nearby contractors. It's surprising, but a lot of them are happy to come help for a reasonable fee. And the cost is well worth it. Probably the person will be very productive, not to mention saving you all sorts of time and materials that you would have wasted on mistakes. Just ask some of the contractors you contacted when you were researching ICF construction in your area.

Remember this saying: "Experienced help: Don't touch foam without it."

Getting a Good ICF Crew

Overview

If you are a general contractor, you will need to find someone to do your ICF work. If you are currently a subcontractor and you want to build ICF walls, you will need to know how to retrain or recruit people for your crew. Neither of these is hard. It boils down to finding people with the right mix of skills and a positive attitude.

And remember, whenever you can, pick people who *want* to try new things. It's a lot easier training workers who are eager to try something like ICFs than it is to push a contractor who is set in his ways.

Finding a Good Crew

For general contractors looking to hire a crew to do their ICF work, there is absolutely no trick to finding a good one. It's the same as finding any other subcontractor. You get names and you check them out.

You can get names in any of the ways this book lists in Chapter 2 for tracking down local ICF contractors. When you locate some people, check with the GCs or owners who hired them. How was their work? Were they responsive? What did they charge?

If you know something about ICF construction, you will better understand what people are telling you. This is one of the reasons it can be useful to get some training even if you aren't going to do the fieldwork yourself. Having researched your local market will help on this, too.

New and Retrained Crews

It is not difficult to hire a few people to create an ICF crew or shift over an existing crew to do ICF work. The key is to start with the right people.

Probably eight out of ten ICF crews are former framing crews. Either the crew used to build wood frame structures and learned ICFs, or the lead person is an experienced framing carpenter who learned ICFs and then hired some people to help him.

This doesn't mean that workers from other trades can't make good ICF crews. Some of the very best were once traditional concrete forming crews that used plywood or steel or aluminum forms. And a few good ones came from masonry or some other trade.

But the fact is that rough carpenters tend to adapt best. About 70 percent of the skills you need to build well with ICFs are things that carpenters already do: layout, cutting to dimension, creating openings, aligning, and squaring. About 30 percent are things that concrete workers already do: setting rebar, specifying concrete, and placing concrete. Carpenters have less to learn—and less to unlearn—than any other trade. Traditional concrete form crews often make the transition well, but it can take more work. Many things that were good enough with heavy, rigid wood and metal forms are no longer quite right with foam forms. And an occasional mason crew can do ICFs well, but for many it's a fight with their training and their nature. John, a supplier in New Jersey, said:

> We have a very successful training program for local contractors. We originally invited masons. It seemed to make sense—this is stacking blocks. But we start teaching them and they pull out their trowels and start wanting to shim things with mortar. It's crazy! No matter what we tell them they want to do it their way. Then they cuss out the product for not working like a concrete block.

It is safest to build your crew around framing carpenters or concrete forms workers. Start with a skilled person to be the supervisor. If this person comes with his own full crew, that should be fine. But he could also add others to build the crew.

Typical Crew Makeup

Lambert, an experienced ICF contractor in Minnesota, describes his rules for an ICF crew:

> The skilled person on the wall crew handles openings, corners, and getting level, plumb, and square. The rest just need to understand the properties of the materials. Just setting the forms is easy. The crew also needs a little concrete experience, which is important for the pour.

Efficient ICF crews usually have three to five people. The lead person is a skilled construction supervisor who is trained in ICF work and, ideally, has a few buildings under his belt. The crew also has one to two other semiskilled workers, preferably experienced framers or forms workers. It also has at least one laborer.

If the lead man is very good, the rest of the crew can have more laborers and fewer experienced workers. The supervisor can direct them, so the other workers just need to have basic reasoning and building skills.

The supervisor assigns tasks, and spends his own time doing or supervising the special tasks that are critical or have tight tolerances. These include building the bucks, forming the walls around openings, determining the positioning of the reinforcing bar, and setting the bracing. The other skilled workers can also perform these tasks, take measurements and cut forms, stack the form walls, and set steel. The laborer can move materials, make cuts and set forms and steel under direction from the skilled crew.

A crew arranged like this can complete all the formwork for one story of a house or small commercial building in 2 to 4 days, with the concrete pour done on the morning of the following day.

If possible, it is ideal to have both experienced frame carpenters and forms workers on the crew. That way you have at least one person highly skilled for every task required for ICF construction. The carpenters can lead the cutting, assembling, aligning, and squaring, and the forms workers can lead the rebar work and the pour. Large crews (over four) should have multiple skilled people. Ideally, those skilled people are about evenly split between frame and concrete workers. However, with ICF training the carpenters can pick up the concrete skills and concrete workers can pick up the carpentry skills.

One disadvantage of having a lot of concrete workers is that it may limit how much of the building the crew can construct efficiently. For some people, a big plus of ICF construction is that the same crew that does the exterior walls can also do any required framing. That means one crew can build the footings (if any), foundation walls, and above-grade exterior walls out of ICFs, and also frame the interior walls, floors, and roof. If the crew includes a lot of concrete workers, they may not be productive when work shifts to the framing.

For commercial projects the concrete work is often more important, and other parts of the building might also be concrete. For this reason, you typically have as many skilled concrete workers as carpenters. In addition, there may be a need for more experienced workers and fewer laborers. A crew of 10 to 12, for example, might only have two laborers.

Experienced Carpenters

Carpenters do well at precision activities like measuring, cutting, and planning. In many cases, a crew made up of carpenters can form footings, build the foundation and above-grade exterior walls with ICFs, then frame the roof, the floor, and all interior walls. This makes it possible to eliminate the separate foundation crew and that reduces time and coordination problems.

But carpenters need to pay more attention to rebar and concrete because they are less familiar with these. They are the structural materials, so they're critical. Installing concrete that doesn't meet strength specs or skimping on the rebar is like using 2×3s instead of 2×4s, or leaving out a bearing wall. It is a huge mistake.

Just as common a mistake is putting in too much of something. It's tempting— when people are confronted with an unclear situation, they assume that more structural material is better, but it could actually *decrease* structural integrity. The design of reinforced concrete *balances* the tensile strength of the steel with the compressive strength of the concrete. Karen, from Nebraska, a structural engineer involved with ICFs for many years, says:

> In their desire to make sure the building is structurally "better" contractors tend to overkill and might even double the steel reinforcement, for instance. This can do more damage than good. When you engineer a building you take into consideration a set of parameters and you optimize them for best performance. If one of them is changed, regardless in what direction, this would make the building behave in a different way from what it was designed to do. Contractors should follow engineer's or generic specifications as closely as possible. Overkill is not always a solution, especially when it comes to steel reinforced concrete and ICFs.

The most critical part of building with ICFs is the concrete pour and the several hours after. During that time all adjustments to get the walls plumb are completed. Most carpenters have limited experience with concrete, scaffolding, and bracing. Crews made up of carpenters especially need to have someone experienced with concrete and ICFs on-site for the pour.

Experienced Concrete Workers

Concrete forming crews are more familiar with rebar and concrete. But they need to pay more attention to the work of assembling the forms. ICFs need to be handled with less force than conventional plywood or metal forms. Foam is great for cutting, forming openings, and making odd shapes, but that is closer to traditional carpentry work, not traditional form setting. And ICFs rely more on a bracing system to hold the walls plumb, straight, and square than conventional forms do.

Experienced concrete workers also have to realize that it takes more time to pour walls, floors, and roofs formed with ICFs. An experienced traditional forms crew can pour the walls for a 1000 square foot basement in $1\frac{1}{2}$ to 2 h. With ICFs, a new crew may take at least twice as long. Experienced crews can work the time down, but they may never get it as fast as it goes with traditional forms, and they should never dump the concrete in one spot or even a few spots around the perimeter. They need to go all the way around and then come back and build the concrete up in lifts.

The Bottom Line

In a growing number of areas of the country, you can find experienced crews that do ICF work on a traditional subcontract basis. They'll quote a price for everything from forming up the walls to placing the concrete and leaving the site clean. You find them and deal with them about the way you do any other sub.

If you want to convert your existing crew to ICF work or hire people one-by-one to build a good crew, that's not difficult. The lead person needs thorough

ICF training. Ideally, the crew has framing experience and some concrete experience. But if the crew has just one of these two, it can train to pick up the other. As you might remember from the last chapter, it is very important for a new crew to get experienced help at the job site on its first project. This help will go a long way to fill any gaps in the crew's skills.

If there is anything else to add, it is that the new crew should *want* to do ICF work. If you have to drag them kicking and screaming, they're probably the type that wants to stick with what they've always done. They might come around at some point, but it's not ideal for them to have to do it on your nickel, and they might never come around. If they are excited about doing something new and better, they're likely to go the extra mile, be flexible, learn what they need to know, and take care of their work. Judging by recent experience, there are plenty of people like this to be found. Contractors who have been through the construction of ICF buildings a few times begin to realize how much better they really are.

Other Subcontractors

Overview

A few trades have to do their work a little differently on an ICF building. The electrician has to adjust the most, but even for him the new skills are few and easy to pick up. The framers (if they're different from the ICF crew), the plumber, the wallboard crew, the finish carpenters, and the exterior finish crew all have to learn to attach their things to ICFs. This is usually very easy to do, but you may have to give them some direction. Ideally you'll also find an HVAC contractor who can size equipment accurately for a super-insulated building. Most of them really can't, and it's worth the time to search for the qualified ones because you stand to save a lot of money and a few headaches if you get this job done right.

Rule number one: Hire those who want to do it

There's a lot to be said for getting subs who are interested and enthusiastic about ICFs and trying something new. They'll pitch right in and get over their small learning curve quickly. They'll probably charge you at a reasonable rate without much negotiating.

However, there are also a lot of contractors who really don't want to be bothered with learning anything new. If you ask a tradesman to quote you on an ICF project and the first words out of his mouth are "That'll never work," or "Whoa! That's gonna cost you," or "I've seen those things before and they're a disaster," then move on. Even these guys usually become reasonable once you show them exactly how easy the work is, and once they learn a couple of simple tricks. But why waste the time and effort? If you really have to use this sub for some reason, then go ahead and educate him. But don't go looking for that kind of grief.

Finding Subs

Since almost any crew in these trades can do their work on ICFs just fine, there aren't many other rules for how to find one. When you interview ICF contractors in your area, you can ask who they use and how those crews worked out. You can also ask the local distributor you're getting your forms from.

You can also ask any crews you currently deal with. If they're reasonable people odds are they'll work out just fine.

Electricians

Electricians need to do a few things differently to get their cable and boxes into ICF walls. Instead of boring holes, pulling cable through, and nailing boxes to studs, they have to cut grooves and rectangles in the foam for mounting the electrical components.

Mark, a builder in Illinois, talks about his first ICF house:

> The "dumb tax" is what I call the extra charge a subcontractor sticks to you when you force him to do something outside his comfort zone. I did this ICF home and the electrician had never installed wiring in an ICF home. He basically doubled his typical bid price and I had no option but to pay. A while after that project was over I met him at a Homebuilders' Association meeting and asked him how it went. He said it was much easier than he was expecting that it didn't take any more time than frame walls. When I asked him for another bid for my next job, he agreed right away to provide one. So I said, "Since you found out it was as fast as with frame, I don't want to pay the dumb tax this time."

Needless to say, it pays to get an electrician experienced with ICFs, or to get a new one up to speed before he bids, if you possibly can.

The key with a new electrician is to demonstrate that the installation is all straightforward. If he uses the right tools, it is just as easy as his usual methods. Thousands of electricians have done it now. If you can have the tools on hand and actually show them how it's done, eight out of ten electricians get comfortable with it right away.

For the cable, the electrician makes grooves in the foam called **chases** (see Fig. 10-1). These go about $1\frac{1}{2}$ to 2 in deep into the foam, depending on local code requirements. The chase should have a bit of a lip on the bottom to hold the cable in place. The recommended tool for the job is usually the hot knife, because it makes a very clean, precise cut. Some contractors prefer a woodworking tool like a router because it moves through the foam faster. The short message is that many tools will do the job. Just drop the cable in the slot. Some like to put a spot of foam adhesive over the cable every couple of feet to secure it.

For the boxes, the electrician cuts out a rectangle (see Fig. 10-2). Some use a large hot knife blade sized to create a snug fit. You can also work a router or other tool around to get the shape. How you secure the box depends on personal preference and what the local inspector requires. Some just fix it to the foam with adhesive. Others cut their boxes in next to a form tie and screw through the side into the tie, just like screwing to a wooden stud. A few put a concrete screw through the back and into the concrete of the wall. That's all there is to it.

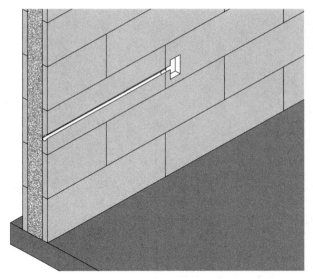

Figure 10-1   Chase and box recess cut into ICF wall.

Framers

If you are hiring a separate framing crew to do interior walls, floors, and maybe a roof, you and your ICF crew should know how these things are going to be attached before the outside walls are built (see Fig. 10-3). It is then easy to explain to the framers.

Figure 10-2   Boxes set in the foam.

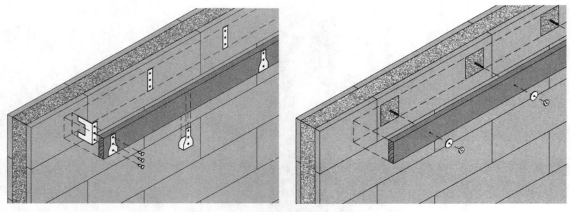

Figure 10-3   Common methods of attaching a floor ledger to an ICF wall.

Floors usually require that some kind of connector be cast into the concrete of the exterior walls, so this is all planned out before the wall ever goes up. There are a half dozen different methods and several different types of connectors to use. Training courses cover it all, and the forms manufacturer can advise you if you're still in doubt.

It's best to consult a new framing crew about the floor attachment in advance, to make sure they understand what they will have to do once they inherit a finished wall with connectors sticking out of it. They may even have some suggestions that will make things easier. It doesn't hurt to have an extra set of eyes on the calculations, to make sure you're casting your connectors in exactly the right place. With concrete the rule is "measure thrice and pour once", because it's a lot easier to do things correctly in advance than it is to fix hardened concrete.

The roof is similar but it is usually a little easier to plan for because there are fewer variables in where the framing members go. The ICF crew embeds structural connectors (anchor bolts or steel straps) in the top of the wall and when everything is cured the framers come in to build and attach a roof (see Fig. 10-4).

Interior walls usually don't require any attachment to the ICF exterior wall. They are supposed to be built in such a way that attachment to the floor below and to the roof members above is enough to hold them rigid. Also, the wallboard usually goes on after the interior walls go up, putting a layer of wallboard (laid against the exterior wall) butting up on each side of the stud (see Fig. 10-5). That tends to hold the interior wall in position.

Sometimes the end stud is tacked to the ICF wall (see Fig. 10-6). Either the local inspector requires it because he's used to seeing it in frame, the customer requires it because he feels that the wall will be more secure, or the builder wants it because he believes it will provide a higher quality product. It is definitely a good idea to anchor a wall that will be subjected to vibration or stress.

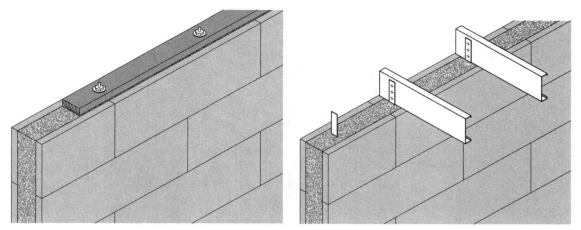

Figure 10-4 Common connectors for attaching a frame roof to an ICF wall.

If there will be a clothes washer or heavy equipment near the frame wall, or a door in the frame wall near the intersection, the joint between the walls can develop drywall cracks. Whether other frame walls (ones that won't be subjected to vibration or stress) need attachment is more a matter of personal preference.

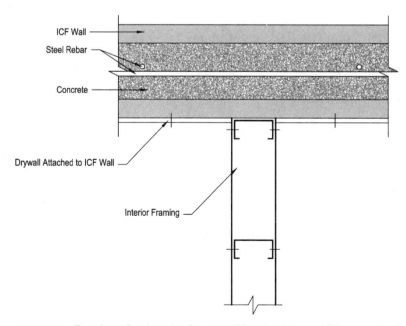

Figure 10-5 Top view of an interior frame wall butting into an ICF exterior wall.

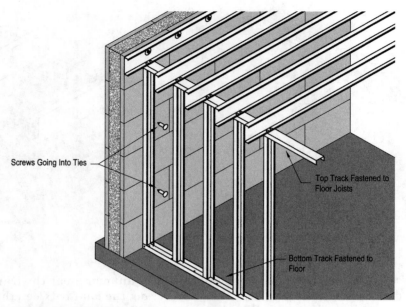

Screws Going Into Ties

Top Track Fastened to
Floor Joists

Bottom Track Fastened to
Floor

Figure 10-6 Attaching the end stud of an interior frame wall to the ICF exterior wall.

In any case, whether and how to connect the walls is planned in advance. There are a dozen ways to make the connection work perfectly well, including screwing into the ties, screwing into steel strapping attached to the ties, and screwing into easy attachment plates set in the wall. The materials are all at hand on the site, or else available from the ICF accessories suppliers.

You may have to watch the price quotes of framing carpenters, too. Duane, a project manager with a homebuilder in Texas, expressed his frustration with pricing from his local framers:

> A framer won't give you any break for having the exterior walls done! Having the exterior walls up saves him time, but they make all their money on the first floor walls. I guess it would save the framer at least one day, depending on the size of his crew.

Residential framers generally quote on labor only since the materials are supplied by the general contractor, and they are accustomed to quoting a price per square foot of the building's floor area. Logically, this should be much less on an ICF building because the exterior walls are already done for them. The exterior walls of most small buildings account for about one-third of total framing materials and labor. So your framer should be quoting a per-square-foot price that is about one-third less than his usual rate, and his total quote should also be one-third less. But a lot of new carpenters just quote their usual rate. This is either because they are concerned that the ICF job may have complications they can't foresee and they want to cover themselves, or they just want to see how much money they can get. You may have to do some explaining and shopping around to other contractors to get a good rate.

This problem usually goes away if you have one crew doing both the ICFs and the framing. Then the crew looks more carefully at what their total labor for all parts will be. And besides, they don't need to pump up their charges so much to get more money because they get paid for the whole building.

Plumbers

Most plumbing is out of exterior walls today, so your plumber is often not an issue. But if the plumber does need to run some pipes in the ICF wall, they go in about the same way that electrical cable does. Just cut a chase for the pipe. If pipe needs to be secured, use standard plumbing brackets fastened to the ties on the side or to the concrete in back with concrete fasteners. If there are through-wall penetrations (hose bibs, sump discharge, etc.), coordinate these as you would with concrete foundation walls.

Wallboarders

Wallboard can go on ICF walls nearly the same as it does on frame. Nails or wallboard screws attach it to the ties. The workers can sight down to locate the ties just like studs. Of course, this assumes that the walls were stacked so that the ties line up reliably. That's usually a given with an experienced ICF crew, but you may want to watch out for this to be sure. If there is much misalignment, you can be sure the wallboard crew will let you know about it.

It's also possible to glue the wallboard in place, using just enough screws to hold it while the adhesive dries. In fact, some contractors feel that this approach can save money and produce a better wall. According to Buddy, an ICF contractor in North Carolina:

> The joints don't have to break on a stud because you have foam behind the wallboard. It's like putting a new layer of board over an old one. You just glue to the foam below and you can break wherever you want. It's faster and there's less waste and when you lean against the board it has solid backing everywhere.

Finish Carpenters

There's not much to adjust in the finish carpentry, either. Baseboards and other horizontal moldings can fasten to the ties, as though they were studs. It's better to use a finish screw because it grips much better than a nail with a smooth shank.

Trim around windows can fasten to the buck that the ICF crew installed into the wall. Other lightweight fixtures (towel racks, toilet paper holders, medicine cabinets, etc.) can attach the same way interior walls do: to the ties, to steel strapping run between the ties, or to the new steel connector plates designed for use with ICFs.

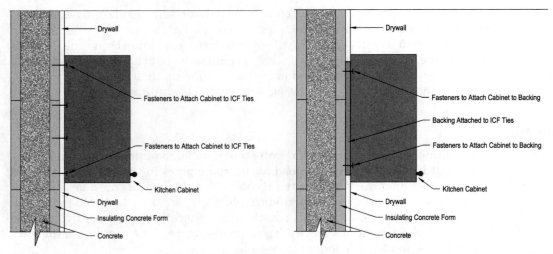

Figure 10-7   Attaching a heavy fixture to the ICF ties (Left) and to a sheet of plywood pre-attached to the ties (Right).

There is more of a decision to be made with very heavy attachments like suspended bookshelves or kitchen cabinets. At one time the preferred method for attaching these was with a layer of plywood. You would decide where these fixtures were going before the wallboard went up. Then you'd put plywood over the wall at this location, screwing it securely to the ties or even anchoring it in the concrete. The plywood needed to be the same thickness as the wallboard. The wallboarders would just butt their board up to the plywood on all sides. Later the fixture would attach to the plywood, covering it up.

But nowadays many ICFs have ties with pullout strengths of well over 100 pounds for a coarse-thread drywall screw. So a lot of contractors just run enough screws into the ties to hold the fixture directly to the wall (see Fig. 10-7). You can consult your forms supplier to decide what to do in your case.

Exterior Finish Crew

Stucco, EIFS, wood siding, vinyl siding, brick, and just about every other common exterior finish goes onto ICFs about the same way it goes onto frame. There are only a few things to learn.

Traditional cement stucco requires the usual steel wire lath. This gets screwed into the ties right over the foam. The foam is a good backer for the stucco. Usually no prep is necessary on the wall. If there are large bumps or gouges in the foam these need to be shaved down or filled in to get a level surface. This is all pretty standard stuff.

An exterior insulation and finish system (EIFS) is popular on ICFs, partly because it usually goes on even less expensively than it does on frame. Over frame a layer of foam board goes on first. But with ICFs the foam is already in place. Again, the crew will have to make sure the foam surface is level, just as on any project. Then the layers of acrylic material and fiberglass mesh go on the way they normally do.

Any siding gets nailed or screwed to the ties as though they were studs. When using nails, make sure they are ring-shank or galvanized. Common nails with a smooth shank will pull out of most ties too easily. Also note that the siding needs to break over a tie to secure the ends of the pieces.

Some people prefer to attach furring strips to the wall, and attach their siding to the strips. This is easy—just screw the strips to the form ties.

Brick or other masonry veneer require a brick ledge, just like on a frame building. The ICF crew builds this into the exterior wall. The brick ties screw into the ICF form ties.

HVAC

The biggest problem with heating and cooling equipment in ICF buildings is that the contractors make it too big. Because ICFs are more energy efficient than frame and most other wall types, ICF buildings don't require as much HVAC capacity to maintain the right temperature.

Still, most HVAC contractors will put the same equipment in an ICF building that they would put into a standard building. There are two reasons for this. The first is that they just don't know what size of equipment they should put in. They don't usually work on highly energy-efficient buildings at all. Normally they just use a rule of thumb—something like "one ton of AC for every 1000 square feet of floor area." and that's all they know. The second is that they don't want to be responsible for undersized equipment. They know that if they put in smaller furnaces and compressors and they are not adequate, then they'll get called back and may have to pay for fixing the problem. They don't know for sure how much they could reduce capacity for an ICF building, so they decide to play it safe by putting in the usual size of equipment.

This results in equipment bigger than it has to be. That means the GC or the buyer pays more than necessary for the building. It's not unusual for HVAC equipment to cost $1000 less in a typical ICF house if the equipment is properly sized. If it's not properly sized, those savings are lost.

It also leads to equipment that short cycles. When the indoor temperature goes up in the summer, the air conditioning comes on and quickly brings the temperature back down because it's so big. The starting and stopping is hard on the equipment and may lead to more frequent maintenance. This is true for the heating as well.

But in the case of the AC, short cycling can also lead to lower indoor comfort. The air conditioning also takes the humidity out of the air. If the AC is only on for short periods it can't take out as much humidity and the indoor air can feel sticky.

Try to find a qualified HVAC contractor and hold onto him when you do. The problem is tough because there are not standard guidelines for how to size the equipment in an ICF building. If there was a simple chart or table that one of the big organizations stood behind, you could hire almost anyone and insist that he stuck to the guidelines. Jeff, an ICF contractor in Colorado, said:

> I have a guy who cuts the equipment in half for an ICF house. He just figures what he would install in a wood frame house and puts in half the size of furnace.

However, there is no guarantee this will be right in other cases. Some contractors use a rule of thumb that equipment can be downsized by 25 percent, but to really be safe, you want someone who can take the specs of a particular building and come up with suitable equipment for it.

You should also be aware that help may be on the way. At the moment we write this, the Portland Cement Association is working on a simple new software package that almost any HVAC contractor can use to get the correct size of equipment to put into an ICF building. The users puts in some standard building specs, and the software gives the equipment specs. If this is available, any reasonably competent and open-minded HVAC contractor might do the job.

In the meantime, your best bets are contractors who have experience with superinsulated buildings or who use modern software to size their equipment. If they've done a number of superinsulated buildings they have more feel for what works. The new software is getting pretty good. Better programs allow the user to enter high R-values for the insulation, high thermal mass, and low air infiltration. They seem to come out with better results for efficient buildings.

If you can't find anyone who you feel confident will do a good job, there is another option. There are a few energy consultants who are familiar with ICFs and will estimate the HVAC requirements of a building from your specs. Some of them have extensive experience and a good track record. You can pay them a fee to do the sizing, then hire a local contractor to install the equipment. They may even help you talk to the contractor if he needs some reassurance. To find an energy consultant, look in the directories of the major web sites about ICFs. They will usually be listed under "Professionals." You may have to sort them out of the listings, however. The Professionals category usually includes a lot of structural engineers and a few other specialists, too. You can go into the ICF discussion forums online and ask which energy consultants other people use.

One additional consideration that the HVAC contractor should be looking at is mechanical ventilation. All types of buildings are being built with tighter and tighter shells to increase energy efficiency and keep out unwanted moisture and pollutants. ICF buildings are usually constructed to be quite tight. ICF walls are about as air-tight as possible with any wall system, and the owners are usually keen on building efficiency so they button up the other parts of the shell, too. But this often means it's a good idea to introduce controlled amounts of fresh air into the HVAC system.

Lon, a longtime ICF general contractor in Oregon, says:

> We always provide a source of fresh air coming into the house. I've listened to some of the other ICF guys talk about how they use air-to-air heat exchangers or energy recovery ventilators. We've never used those, but we bring fresh air into the air

system some way to get whole house ventilation. Nobody requires us to do it, and I know some other people don't and I haven't heard of terrible problems if they don't. But we think it's a good idea. Even some stick framers are starting to do it.

In practice, about half of the ICF houses constructed today have special ventilation measures and half don't. Most people agree it's a good idea for all buildings, and eventually regulation is likely to require it. The American Society of Heating, Refrigeration, and Air Conditioning Engineers (ASHRAE) has tentatively agreed on a new standard that includes provisions for bringing fresh air into almost all buildings. When this is finalized, it could well be adopted by the building codes.

But in any case you want an HVAC contractor who knows something about these issues and can advise you. Bringing in fresh air doesn't have to be expensive, but someone has to figure out whether it's necessary and how to size it and install it. Fortunately, the contractors who are equipped to deal with high-efficiency buildings also know something about fresh air ventilation. The two things are related, so they commonly come up in the same buildings and the same people know about them. If you can find someone competent with one, you've usually found someone competent with both.

How Commercial Is Different

Finding and managing subcontractors in commercial construction can be a lot different. Mostly it's easier. In commercial construction, the subs are usually more used to working with concrete. The HVAC systems are often designed by mechanical engineers. They are more likely to know what to do with ICF walls and have all the skills and tools.

The floors, roofs, and interior walls may not be simple frame. Commercial construction involves a lot more heavy steel and concrete members. But the crews have generally connected these things to cast concrete walls, so they won't often need to be told how to do it. If it is a little different because of the foam they can probably figure it out. Most commercial buildings are professionally designed, anyway. You may have to educate the architect and engineer about ICFs, but they should be able to run with the system once they understand it.

11

Estimating

Overview

Some contractors say that when they started with ICFs they actually had more trouble with estimating than construction. It is hard to know exactly how much it will cost to do something you've never tried before, and if you're wrong it can end up costing you a lot of money.

Early on, you're also shooting at a moving target. Your later jobs will have substantially less waste and take less labor than your first one.

Some manufacturers offer estimation software that can do a lot to organize the problem for you, do calculation, and tally up the results. Otherwise, estimation with ICFs is the same as it is with anything else. You need to list all the materials and the labor tasks, work out how much you will need of each, price them, then multiply and add.

If You Are the General Contractor

If you're hiring someone else to do the ICF work, of course you don't have to estimate. You ask the sub to do that and give you a total cost. Like anything, the number you get can vary a lot depending on the project and the sub. Typical rates for completed ICF walls are $6.50 to $8.50 per square foot of gross wall area, but if you don't know what this means, it can be more dangerous to use than useful.

The ICF quotes are almost always for *wall area*. This is different from, say, framing quotes, which can be figured on the square footage of the *floor area*. Floor area is a lousy way to figure ICF costs because those costs get bigger when the walls get bigger, not when the floor area does. For example, if you change the design of a house so that the exterior walls go from 8 to 10 ft tall, the exterior wall area and the cost of building the walls out of ICFs go up by about 25 percent, even though the floor area stays the same.

Gross wall area is the total area of the ICF walls, not taking out anything for openings like windows and doors. You don't usually take out for openings because what they save in materials goes into paying for the extra labor to create them. If you have a really big opening, like a picture window so big that it takes up most of one wall, then this rule doesn't apply anymore. Then the quotes may well come in lower because a lot less material is used.

The low end of the price range applies mostly to simple basements. These are big rectangles with only four corners and few openings. Setting the forms on this type of project is very fast and there is almost no waste. Typical above-grade walls run around $7.00–$8.00. When you get into above-grade walls with more corners, lots of openings, and maybe some odd corners, you'll probably end up around $8.00. In something really elaborate with curves, overhangs, and such, expect even more.

Materials

The major materials used in ICF wall construction are as follows:

- The forms
- Concrete
- Buck material
- Rebar
- Foam adhesive

These are more or less in order of most cost to least cost. Buck material is a lot less in basements or other structures without many openings. The amount of rebar used depends a lot on the structural loads of the building and the local area. Some contractors use other things instead of adhesive.

There are also minor items like tie wires or cable ties to hold rebar, screws, and so on. So there are really only five materials that have to be accurately estimated. Forms are currently running around $3.00 per square foot, but get a direct quote from your supplier. There's always variation. Also, the special forms (corners, brick ledge, etc.) are usually more, so if you have a lot of them the cost for your project will run higher.

Concrete is usually pretty easy to estimate. The local ready-mix supplier should give you a quote per yard from the specs you give. Most manufacturers have simple rules for figuring the concrete needed. For example, with some forms it takes one yard of concrete to fill 13 blocks. Be sure you're dealing with actual blocks in the wall when you use this kind of method, not including waste blocks.

There are few shortcuts with buck material and rebar. You just have to look at the dimensions and add them up. For rebar, it can be handy to take elevations of the plans and draw in all the bars with a red pencil. That will help you see all the pieces and the overlaps you need. Then you can just count.

Adhesive is very individual. Just ask another contractor how much he uses on a typical job.

A tricky part of materials estimation is the waste factor to use. That's one part of the target that moves a lot. Very experienced ICF crews report waste as low as 2 percent on forms and concrete, but on the first job close to 10 percent is more common. After the first job, 5 percent is a good rule until you get more experience. And you probably never want to cut it too fine on either of these things. There are few experiences more frustrating than standing on the scaffolding and finding you are six blocks short of completing the form walls, or half a yard of concrete short of filling your walls.

It is worth a few words here to warn you about running short of concrete. It's more than just frustrating. If you stop filling a wall and come back to it a few hours or a day later, you will get a **cold joint**. This is a sort of seam in the concrete. The first batch of concrete you placed will be cured enough by the time you put the rest on top of it, that the two batches won't bind well to each other. This creates a weak spot in the wall. This could be a big no-no depending on where it occurs, what the loads on the wall are, and what the rebar pattern is. So when it comes to cold joints, the rule is *don't have them*, even if it means ordering a little extra concrete to be sure. Of course, between stories of a building it's usually expected you'll have a cold joint and the rebar is designed to take care of that, so no worries there.

Waste on buck material can be high because you have a lot of short pieces to cut out of fixed lengths of material. Some of the plastic buck material can be custom cut, but if you use lumber you pick from the standard lengths. Rebar can usually be figured pretty fine if you have taken account of the overlaps in your estimate. You usually order the bars to custom lengths, so there are few cutoffs that can't be used.

Labor

Labor rates can vary a lot depending on the experience of the crew and the complexity of the job. The hours to build an ICF wall can drop by 30 to 50 percent from a crew's first job to its fifth. Once the crew is experienced, a set of typical above-grade walls (medium complexity) can usually get built with about one worker-hour for every 20 square feet of gross wall area. This includes everything from laying out the walls to cleaning up the site after the pour.

But on a very simple basement an experienced crew might make, say, 35 or 40 square feet per worker hour. In a highly cut-up building, the rate might drop to 15 square feet per worker hour or less. In addition, some contractors claim they can get better productivity with some ICF systems than others. You'll have to talk to them and check your own production rates for precise numbers. The products are constantly changing to incorporate new improvements, so it's hard to generalize.

Like anything, the best estimate of labor hours comes from past experience. Since you won't have any on your first ICF job, you will need to rely on the advice of other contractors and your ICF supplier. You might add a little cushion to make sure.

You might also want to make sure that you are well trained. Especially on your first job, knowing what to expect and how to handle common situations will save a lot of time in head scratching. Simply put, training trims time.

Once you have a few jobs under your belt, your judgment will provide the best estimate of labor productivity for your next job.

Equipment

Most contractors add the cost of tools and equipment in their labor rates, unless they have some especially expensive equipment. Most of the things necessary for ICF construction are already in the carpenter's toolbox. And many of the rest are small, inexpensive items.

There are two exceptions. One is the bracing/scaffolding system. A complete setup can cost a few thousand dollars. Figure this in the same way you figure any other large equipment like your trucks and lifts. If you rent a system, get a quote on that directly. If you like figuring it into your labor hours that's fine, too. If you like a separate charge, estimate how long you think the equipment will last and charge a fraction of the total cost to each job.

The other exception is the concrete pump. Many contractors prefer to use a pump even when they could probably pour directly from the chute of the ready-mix truck. The pump is much more precise and easy to control. For the cost of the pump, contact the local supplier. A half-day rental is almost always enough for a pour if you plan it well. A boom pump typically rents for over $500 for a half day, and that includes the operator. A line pump usually rents for under $500, but you need to have extra crew on-hand to move the heavy line around.

Software

One of the handier things some of the ICF suppliers offer is estimating software. This is usually in the form of a spreadsheet that runs on Microsoft's Excel program. It is of course set up to estimate costs only for the supplier's own brand of ICF. This may seem limiting, but it actually lets the spreadsheet do more of the work for you. Since the program knows what system you're using, it can provide information like the size of the form, the amount of concrete that goes into each form, and so on. Some of the spreadsheets allow you to upload a price list for all the parts of the ICF system, and include tables of estimates for typical man-hour rates.

Of course, if you're handy with computers you can always make your own spreadsheet. That makes sense after you have done a few jobs and get an idea of exactly what you want the spreadsheet to do (see Fig. 11-1).

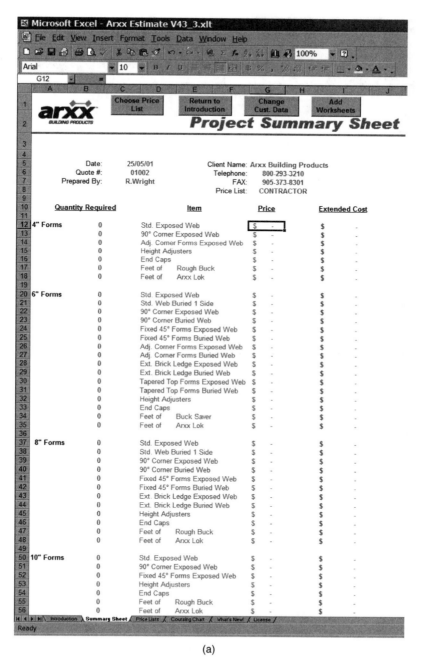

(a)

Figure 11-1 Views of estimation spreadsheets (a–f) available from ICF suppliers.

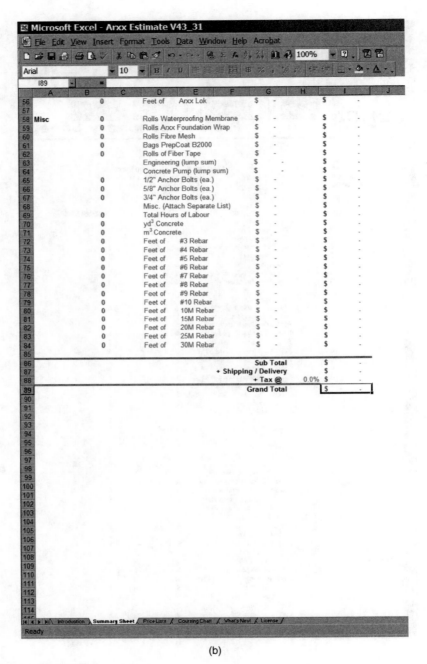

(b)

Figure 11-1 *(Continued)*

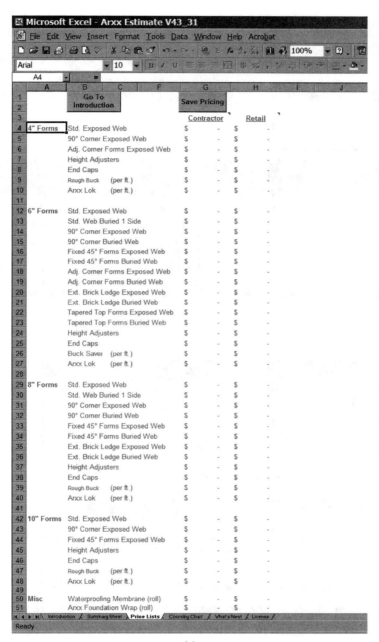

(c)

Figure 11-1 (*Continued*)

Microsoft Excel - rewardestimating1_1.xls

File Edit View Insert Format Tools Data Window Help

Arial   ▼ 10 ▼ B I U

E22   =

Customer Information	
Customer	Customer Name
Address	Customer Address
City, State, Zip	Customer City, State, Zip
Phone	Customer Phone Number
Contact Person	Customer Contact

REWARD. WALL SYSTEMS

iForm   eForm

Wall Information

Wall #	Wall Height	Linear Feet	# 90's	# 45's	Form Used	Vert. Rebar Spacing	Horiz. Rebar Spacing	# Courses	Vert. Size	Horiz. Size	Top Plate (Y/N)	Ledger Board (Y/N)	Brace (Y/N)	Ledge Form (Y/N)
1														
2														
3														
4														
5														
6														
7														
8														
9														
10														

Conversion Ex. 4'-9" = 4.75	
Inch	Decimal
1	0.08
2	0.17
3	0.25
4	0.33
5	0.42
6	0.50
7	0.58
8	0.66
9	0.75
10	0.83
11	0.92

Form Used	Window Sizes			Form Used	Window Sizes			Form Used	Door Sizes		Qty
	Width	Height	Qty		Width	Height	Qty		Width	Height	

Other Project Information	
Number of Stories, Including Basement	
Number of Angles, Excluding 90's	
Number of T-Walls or Pilasters	
Number of Hours the Pump Will be There	
Number of Trips the Pump Will Make	
Will Address Above be Shipping Add. (Y/N)	
Number of Crew-Members	
Will You Incur Lodging Expenses (Y/N)	

Shipping Information	
If "N" was selected at left.	
Name:	
Address:	
City,St,Zip:	
Phone:	

Instructions \ Enter Project Info Here \ Man Hour Factors / Estimate - Summary / Estimate - Itemized / Company Admin / Prices /

Ready

(d)

Figure 11-1 *(Continued)*

Price List

Enter Cost For Items Listed Below

Description	Unit	Price
9 1/4" e-Form Straight	eForm	$ -
9 1/4" e-Form 90 Degree Corner	eForm	$ -
9 1/4" e-Form 45 Degree Corner	eForm	$ -
11" e-Form Straight	eForm	$ -
11" e-Form 90 Degree Corner	eForm	$ -
9" I-Form Straight	iForm	$ -
9" I-Form 90 Degree Corner	iForm	$ -
9" I-Form 45 Degree Corner	iForm	$ -
11" I-Form Straight	iForm	$ -
11" I-Form 90 Degree Corner	iForm	$ -
11" I-Form 45 Degree Corner	iForm	$ -
13" I-Form Straight	iForm	$ -
13" I-Form 90 Degree Corner	iForm	$ -
13" I-Form 45 Degree Corner	iForm	$ -
11" I-Form Ledge	iForm	$ -
13" I-Form Ledge	iForm	$ -
Concrete	Cu. Yds.	$ -
# 4 Rebar	Lin. Feet	
# 5 Rebar	Lin. Feet	$ -
# 6 Rebar	Lin. Feet	$ -
# 7 Rebar	Lin. Feet	$ -
# 8 Rebar	Lin. Feet	$ -
Ledge Form Rebar	Piece	$ -
Foam Adhesive	Cans	
Foam Adhesive Cleaner	Cans	$ -
Window Buck Material	Lin. Feet	$ -
Window/Door Buck Bottom Material/Bracing	Lin. Feet	$ -
Door Buck Material	Lin. Feet	$ -
Anchor Tunnel	Pkg 100	$ -
1/2" Anchor Bolts	Each	$ -
rBase - Bracing/Scaffold/Alignment System	Each	$ -
Misc. Wire, Fasteners,OSB, Etc.	Factor	$ 0.15
Enerfoam Gun	Each	$ -
Amount of Shipping	Total Amt	$ -
Concrete Pump Hourly Rate	Hr Rate	$ -
Concrete Pump Trip Rate	Trip Rate	$ -
Daily Travel Rate	Daily Rate	$ -

(e)

Figure 11-1 *(Continued)*

You can not enter data on this page. All values are computed by the program. See instructions.

REWARD.
WALL SYSTEMS

Your Company
Your Address   Your City, State, Zip
Your Phone Number   Your Web Address   Your Email Address

Name:	Customer Name		Est. Days to Complete	#DIV/0!
Address:	Customer Address		Installed $ per Form	#DIV/0!
City, State, Zip:	Customer City, State, Zip		Installed $ per Sq. Ft.	#DIV/0!
Phone:	Customer Phone Number			

Category	Description		Quantity	Unit Price		Ext Total	
9.25 eForm	9 1/4" Straight	eForm		$	-	$	-
9.25 eForm	9 1/4" 90 Degree Corner	eForm		$	-	$	-
9.25 eForm	9 1/4" 45 Degree Corner	eForm		$	-	$	-
11 eForm	11" Straight	eForm		$	-	$	-
11 eForm	11" 90 Degree Corner	eForm		$	-	$	-
9 iForm	9" Straight	iForm		$	-	$	-
9 iForm	9" 90 Degree Corner	iForm		$	-	$	-
9 iForm	9" 45 Degree Corner	iForm		$	-	$	-
11 iForm	11" Straight	iForm		$	-	$	-
11 iForm	11" 90 Degree Corner	'Form		$	-	$	-
11 iForm	11" 45 Degree Corner	iForm		$	-	$	-
13 iForm	13" Straight	iForm		$	-	$	-
13 iForm	13" 90 Degree Corner	iForm		$	-	$	-
13 iForm	13" 45 Degree Corner	iForm		$	-	$	-
11 Ledge	11" Ledge Form	i Form		$	-	$	-
13 Ledge	13" Ledge Form	i Form		$	-	$	-
Concrete	Concrete	Cu. Yds.	2	$	-	$	-
Rebar	# 4 Rebar	Lin. Feet		$	-	$	-
Rebar	# 5 Rebar	Lin. Feet		$	-	$	-
Rebar	# 6 Rebar	Lin. Feet		$	-	$	-
Rebar	# 7 Rebar	Lin. Feet		$	-	$	-
Rebar	# 8 Rebar	Lin. Feet		$	-	$	-
Rebar	Ledge Form Rebar	Piece		$	-	$	-
Foam	Foam Adhesive	Cans		$	-	$	-
Foam	Foam Adhesive Cleaner	Cans	1	$	-	$	-
Bucks	Window Bucks (Top & Sides)	Lin. Feet		$	-	$	-
Bucks	Window Bucks Bottom/Bracing	Lin. Feet		$	-	$	-
Bucks	Door Bucks (Top & Sides)	Lin. Feet		$	-	$	-
Bucks	Door Bucks Bottom/Bracing	Lin. Feet		$	-	$	-
Anch. Tunnel	Anchor Tunnel	Each		$	-	$	-
Anch. Bolts	Anchor Bolts	Each		$	-	$	-
rBase	Bracing/Scaffold/Alignment Sys.	Each		$	-	$	-
Misc.	Wire,Ties,Fasteners,OSB, Etc.	Factor		$	0.15	$	-
Pump	Concrete Pump Hours			$	-	$	-
Pump	Concrete Pump Trips			$	-	$	-
Shipping	Shipping to Job Site					$	-
Material & Equipment Sub Total						$	-
Tax						$	-
Mat. & Eq.	Total Projected Costs of Materials & Equipment					$	-
Labor	Labor - Rebar, Forms, Bucks, and Concrete			$	-	$	-
Travel	Hotel and Travel Expenses						-
Labor Sub Total						$	-
Tax on Labor						$	-
Labor	Total Projected Costs of Labor					$	-
Overhead						$	-
Profit						$	-
Total	Total Estimate Amount					$	-

(f)

Figure 11-1   *(Continued)*

12

Special Construction Awards Programs

Overview

There are now several programs that give financial incentives or marketing recognition to buildings with superior construction. ICF contractors can use these programs to help market their work. The buyers can use them to get cold hard cash from their ICF building.

Many of these awards come fairly easily to ICF structures. If you're building with ICFs it's worth looking into them and the marketing advantages they can give you.

There are three types of useful programs: energy efficiency, green construction, and insurance classifications. Many different organizations will rate the energy efficiency of a building. Mostly these programs are for single-family homes, but a few will rate apartments or other buildings. The builder has to sign up for the program and cooperate with the officials who will do the rating. If the building gets a high energy-efficiency rating, the builder can advertise this to buyers and possibly get more market interest and a premium price. But there's more to the energy ratings than just the direct marketing value. Some mortgage lenders will give the buyer better terms if a house has received a high-energy efficiency score from a major program. This means the buyer may be willing to pay more for the house because the mortgage can be more favorable. The builder gets more for the house, or the buyer pays less for the mortgage, or each one gets a little of both. In addition, some lenders will allow the buyer to borrow more than that borrower would normally qualify for. So a lot of people who otherwise wouldn't be able to consider the house you built could now afford it. The pool of prospective buyers is bigger, and the house becomes easier to sell.

The energy-efficient mortgages can get a bit complex, but if you go through the specifics in this chapter step by step you should be in good shape to understand them and show buyers how they can get one. They also involve some paperwork and procedures, but this is also manageable if you know what you're doing.

Green construction ratings are another kind of stamp of approval you can get on a building. Independent rating organizations score buildings based on the environmental friendliness of their features—materials, design, and so on. The builder can advertise the rating and possibly get more interest from buyers. Even more important, some buyers and designers now require high scores as part of their bid requirements, and are looking for builders who can comply.

Most insurance companies have special classifications for houses or commercial buildings that have superior construction. These are buildings that suffer less damage in fires or natural disasters. When a building gets this kind of classification the insurance premiums are lower, so the owner saves money. But on top of the savings, when the insurance agent treats your building this way, it gives the buyer a third-party confirmation that ICFs are superior construction. Whether they care more about the money or the stamp of approval, this is one more benefit of an ICF building that you can use in the sales pitch.

The Story of Energy Efficiency Ratings and Stretch Mortgages

Since the energy crises of the 1970s, there has been a lot of pressure on all parts of society to encourage people to save energy. One result of this has been that banks came up with something called an **energy efficient mortgage** (EEM). With an EEM, homeowners buying very energy efficient homes get favorable terms, like the right to borrow more money or lower interest rates. There is a variation on this idea called an **energy improvement mortgage**. With an energy improvement mortgage buyers are entitled to extra financing to cover the cost of upgrades that increase the home's energy efficiency. The loan requirements and loan limits of the two programs are figured differently, but it boils down to about the same thing from the borrower's point of view.

The logic of the energy efficient mortgage is something like this. Banks have limits on how much money they will lend a buyer for a new house. This is because they have found that if they lend too much, the monthly payment will be too high, the buyer won't be able to cover it, and the buyer stands a good chance of defaulting on the payments. The bank has to repossess the house and try to resell it. This is unpleasant, it's costly, and depending on how much the bank lent it might not get all its money back. But, if the house is very energy efficient, the monthly utility bills will be less. The owner will save money on energy, which is then available to help pay the mortgage. So the bank figures it can lend the buyer a little more and have no greater chance of default than for the typical customer in an ordinary house.

For example, a particular lender might normally give someone a mortgage with a monthly payment equal to 28 percent of the person's income. The lender has found that if you keep the payment at or below 28 percent, people tend to make their payments. So, given this person's income and the current interest rate, our borrower might be able to make the payment on, say, a $250,000 house. But, if the house is especially energy efficient, the fuel bill savings could be substantial. In that case, the lender could justify giving the buyer a mortgage with a payment that is a bigger share of his income. The buyer might be able to buy a house with a price tag more like $275,000. And he is no more likely to default because his total expenses are no worse than they would be if he bought a conventional house for $250,000. He will simply pay less to the utility and that much more to the lender. His total monthly housing expenses are the same.

This type of program has several benefits for the builder of an ICF house. First, more people can afford your house. People with an income just below what's needed to get a mortgage for the house will now be over the minimum. And with more potential buyers the house can move faster.

Second, you can go ahead and charge more for the energy efficiency features. The buyer is willing to pay more because the higher mortgage covers the cost and the lower utility payments compensate for the higher mortgage.

Third, the program is an independent verification of the value of the house. A third party rates the house as energy efficient, and the lender is willing to stake its money on that rating by giving the buyer a bigger mortgage. This is a strong testimonial to show to buyers. It's clearly *not* just some sales gimmick—a trained technician comes out and makes the measurements on the house.

Some lenders offer additional benefits beyond just giving the borrower a bigger loan amount. Using the same logic of lower utility bills, they might offer a lower interest rate. Or they might offer to lend an amount that is a higher percentage of the total purchase price (for example, 95 percent of the total house price instead of 90 percent). Or they might offer lower closing costs, or some combination of these different benefits.

Who's involved

There are three separate groups that do things related to energy efficient mortgages, but you may not need to get all of them involved in every case. These groups are the energy raters, the federal government energy programs, and the mortgage lenders.

The energy raters are paid professional service companies that inspect new houses to verify that they meet energy-efficiency guidelines. For almost any energy efficient mortgage you will have to hire one of these to inspect your house.

The federal government energy-efficiency programs in the United States and Canada are really more marketing and promotion services than anything else. They have a brand name and a logo and big web sites, and they let builders who meet certain energy efficiency standards and lenders who offer energy efficient mortgages use the name and the logo and list themselves on the web site. They put a big stamp of "approved" on businesses that get involved in the energy efficiency programs.

The mortgage lenders give the homebuyers the preferential loans and rates. But which lenders do it and how you find them is not at all standardized. So getting the mortgage itself can take some research. But the application process is pretty similar from one lender to the next. Generally speaking, a lender will give the favorable mortgage if it gets paperwork to show that the house has been inspected by a certified rater and received a high energy efficiency rating. Some may require that the builder or the house also be registered with the federal energy efficiency program, but usually an acceptable score from the rater is the main requirement.

Once you get past the initial learning hurdles, the energy efficient mortgages can be a big boon. Your houses look that much better to the buyer. More people can afford them. You can charge more for them. And the occupants may be happier with the results when their bills each month remind them that they got a break for having a superior building.

So read about energy efficiency programs here, and keep your eyes peeled for new developments. Consider getting involved in one of these so your houses can be rated. And when you do, try to find out what lenders give favorable terms so you can pass these names on to prospective buyers. With some people, the option to get better terms on the mortgage will be a big selling point.

Home Energy Rating Services

The way things have evolved in the United States, the official raters of energy efficiency are the **home energy rating service (HERS) providers**. Some of these are state-sponsored organizations and some are private corporations. But all are licensed and approved to perform inspections of houses for energy efficiency. Anybody can do tests and estimate how energy efficient a house is. But virtually every lender that offers an energy efficient mortgage and every government program that offers promotional benefits to efficient houses relies on the ratings from the HERS providers to determine whether a particular house qualifies for their programs or not.

There are over 60 HERS providers in the United States. They cover nearly the entire country. You can find them in a directory on *www.energystar.gov*, the web site of the U.S. government's energy efficiency program.

In Canada, the system is a bit different. You can start by going to the sites covering Canada's R2000 program, *www.chba.ca/r2000* and *www.r2000.org*.

HERS ratings are done on a standardized scale. They assign each house a number from 0 to 100. A rating of 80 equals the energy efficiency of a house designed to the minimum standards of the 1993 Model Energy Code (MEC). A house with an estimated 5 percent lower energy consumption than a comparable MEC-minimum house gets a rating of 81. A house with an estimated 10 percent lower energy consumption gets 82, and so on.

Most of the houses standing in the United States were built in past years according to lower energy standards. To get some sense of comparison, they have estimated energy efficiencies more in the 30 to 75 range. This corresponds to energy consumptions of about 30 to 150 percent *higher* than MEC-level construction.

You can hire a HERS to rate your house for any reason you want. And if you do it just for yourself, you can have them do it any way you want. But there is a standard HERS energy estimation procedure, and if you want to use the results to qualify for a particular program or energy efficient mortgage, the rater will have to follow the standard procedure.

The standard HERS rating procedure focuses on a few key aspects of the building:

- Tightness of the construction
- Insulation
- Tightness of the ductwork
- Energy performance of the windows
- Efficiency of the HVAC

If you build with ICFs, you'll automatically have a tighter building and much superior wall insulation, compared to standard home construction. To match you in those areas, a stick builder will usually have to spend hundreds or thousands of dollars for added measures on his walls. So you'll have a head start. Just how good a rating you end up with will depend on what you put into the rest of the shell (especially the roof) and the other three categories.

The HERS rating procedure involves two steps—a plan review and on-site testing. For the plan review the builder has to provide the following:

- Construction drawings, including details of insulation levels and placement
- Specifications for windows
- Specifications for space conditioning and water heating equipment
- Procedures to be used for duct sealing
- Procedures to be used for sealing the building envelope

The rater will examine this information and give a preliminary estimate of the house's rating. If the design changes, the builder needs to tell the rater so the rater can take the changes into account and come up with a new preliminary estimate.

Once the house is constructed, the HERS rater does the on-site inspection. That includes the following:

- Checking to make sure the energy-efficiency features shown in the plans are in place
- Air leakage testing
- Duct leakage testing
- Collection of the house's dimensions

If things are as promised on the plans and the leakage tests are good, the rating becomes final. You get paperwork that certifies the score of the house. You can show it to buyers or use it to get the house into an energy-efficiency program or qualify it for a special mortgage.

Energy Star

The federal energy-efficiency marketing program in the United States is called **Energy Star** (see Fig. 12-1). The U.S. Department of Energy actually runs several different programs under that name for the promotion of energy efficient appliances, new houses, building renovations, and so on. But the new house program is the one of interest to us, and it has become quite well known and useful to the energy efficient builder and to the consumer.

In Energy Star's home program, a builder can get a house rated and, if it meets required guidelines, the builder will get official documents that certify it is an Energy Star house. Then he can use the Energy Star logo in marketing, tell people the house is an Energy Star house, and so on. And some lenders will also accept the Energy Star rating for their energy-efficient mortgage programs. So if the house gets an Energy Star certification, the testing part of the application for their EEM is done.

A lot of information about the program is available on the web site *www. energystar.gov*. Almost any single- or multifamily residential construction under three stories tall can be included. The best way to start is to become an *Energy Star Partner* by signing the partnership agreement, which is on the web site. Then you need to arrange to get your houses rated.

There's more than one way to get a house rated for Energy Star. But all are designed to verify that the house is at least 30 percent more energy efficient than it would be if it were built to the minimum standards of the 1993 MEC.

The simplest way to have a house rated is to submit it to standard HERS testing and get a score of 86 or above. (An 86 corresponds to 30 percent more efficient than MEC-minimum.) Your HERS will tell you the preliminary rating based on your house design. If it is below 86 he can also give you advice for things to change to raise the energy efficiency. You are then free to change the plans and resubmit them. Of course, the actual construction has to match the plans when the rater comes out to inspect it.

The other procedure to have a house rated for Energy Star is with **builder option packages** (BOP) (see Fig. 12-2). With this method the cost of the rating can be less, but it has some restrictions.

With the BOP method there is no plan review, and that lowers the fee to the rater. Instead, the builder designs the house and picks the components that will go into it according to a detailed list of specs that cover everything from insulation material to wall thickness. There are many different lists of specs you can use, and each one is called a builder option package. There are BOPs that include ICF walls, and of course houses with ICF walls don't have to have so much of some other features.

The BOPs are preverified for energy savings. In other words, tests show that a house built according to a BOP will have a HERS rating of 86 or above, so there is no need for the rater to check out the plans again. The rater performs the on-site inspection to make sure that everything specified in the BOP is in place, and the tests meet requirements. The house gets a *pass* or *fail*, not a numerical score. The HERS professionals are almost all qualified to do a BOP rating, so by and large you contact these same people to do the work.

Will Your New Home *Really* be Energy-Efficient?

Ask Your Builder These Key Questions:

☐ **Has the home qualified for the ENERGY STAR label?**
The ENERGY STAR label assures you that your home's predicted heating, cooling and hot water energy use is at least 30% less than a comparable home based on the national Model Energy Code. Be aware that outfitting your home with ENERGY STAR labeled products (e.g. windows, lighting, appliances, etc.) will not necessarily make it an ENERGY STAR labeled home.

☐ **Are the home's windows appropriate for the climate?**
With improvements made in window technologies, it is now possible and more affordable to buy energy-efficient windows designed for your home's specific climate. Houses in colder climates should have windows with a low U-value, effectively holding heat in the house and preventing condensation. In hot climates, a low Solar Heat Gain Coefficient (SHGC) is important, allowing visible light into the house while blocking out heat. Look for a window's National Fenestration Rating Council (NFRC) rating to find these specifications, or simply look for windows with the ENERGY STAR label.

An ENERGY STAR® Labeled Home

☐ **Is the home's insulation optimized and was it properly installed?**
Insulation in a home's walls and attic serves as a protective barrier, keeping out excessive heat and cold and maintaining even temperatures between and across rooms. The effectiveness of insulation increases as its R-value increases. For insulation to work properly, it also *must* be installed carefully, without gaps, crimping, or compression. This is most important in places where insulation fits around pipes, electrical wiring and outlets, or other obstacles.

☐ **Is the home's building envelope properly sealed and tested for air leakage?**
The average home has hundreds--if not thousands--of small holes through which heated or cooled air escapes to the outside. Those holes also allow moisture, dust, pollen, and insects to enter your home. A tightly sealed and properly ventilated home, verified on-site by a home energy rater, will not only reduce your energy bills but also improve your home's indoor air quality.

☐ **Is the home's heating and cooling (HVAC) equipment highly efficient and properly sized?**
Furnaces rated at least 90 AFUE and air conditioners rated 12 SEER or higher qualify for the ENERGY STAR label, meaning they are very energy-efficient. This equipment saves you money, often comes with longer warranties, and requires less maintenance. Also, when a home is built with energy-efficient windows, optimal insulation, and tight (not leaky) construction, smaller HVAC equipment can more effectively maintain comfort and will last longer.

☐ **Is the home's ductwork tightly sealed, sufficiently insulated, and tested for air leakage?** Tightly sealed ducts are crucial to energy efficiency. Typical ducts leak 20-30% of the air forced through them, wasting 20-30% of the money you spend on heating and cooling. With proper sealing and insulation, verified on-site by a home energy rater, you can substantially reduce these leaks.

Tight Construction

Tight Ducts

Improved Insulation

Energy-Efficient Heating and Cooling Equipment

High Performance Windows

For more information on ENERGY STAR labeled homes, or to find an ENERGY STAR partnered homebuilder or energy rater, visit our Web site at www.energystar.gov or call toll-free 1-888-STAR-YES (1-888-782-7937).

Figure 12-1 Checklist from Energy Star of features for the homeowner to look for.

Instructions for Using ENERGY STAR® Builder Option Packages

Builder Option Packages (BOPs) are a prescriptive method for labeling new homes ENERGY STAR. BOPs specify levels and limitations for the thermal envelope (insulation and windows), HVAC and water heating equipment efficiencies for a specific climate zone. BOPs require a third-party verification, including testing the leakage of the envelope and duct system, to ensure the requirements have been met. Follow these steps to build an ENERGY STAR labeled home using a BOP:

1. To find the BOP, visit the ENERGY STAR Web site at www.energystar.gov/homes. Choose the "Resources" link and click on "Builder Option Packages" under the "Support Resources" section.

2. Choose the state and county where the home will be built, and open the File. Opening the BOP files requires Adobe Acrobat Reader; a free version of Adobe Acrobat Reader can be downloaded from www.adobe.com.

3. Identify the package (i.e., BOP Number) that you are interested in building. There may be more than one page of BOPs to choose from, depending on your location. Make sure that the house you are building meets the limitations of the package. For example, if the prospective home has 16% window area, the BOP selected must meet or exceed the corresponding limitation - i.e., chose a BOP that allows </= 18% or 21% window area.

4. Build the home, following all the BOP specifications. For clarification on certain items please read the attached "Footnotes" section.

5. Contact a BOP provider to get your home inspected and labeled ENERGY STAR. BOP providers can be located on the Locator Map of the ENERGY STAR Web site at www.energystar.gov/homes.

6. The BOP provider will send a BOP inspector to verify the home meets or exceeds all requirements listed in the BOP. Verification of the home typically includes testing the air leakage of the envelope and duct system. If the home complies with the BOP, the inspector will sign and date the BOP sheet. This sheet is then filed with the BOP Providers for their records.

7. For home buyers interested in an ENERGY STAR mortgage, Fannie Mae requires estimated monthly energy cost savings. For BOPs, these estimates are determined using the monthly cost savings table developed for each climate zone, such as the table below. To use this table:

- Choose the appropriate number of stories, foundation type, and home size that most closely fits the home being built and locate the estimated monthly savings.
- Insert the estimated monthly cost savings in the appropriate line at the bottom of the BOP sheet. Note that these estimated savings should NOT be used as basis for guaranteeing utility bills. This should only be done on a case by case basis with a qualified energy modeling tool.
- Submit a copy of the signed BOP, which includes the estimated monthly cost savings, with your loan request forms, and indicate your interest in receiving an ENERGY STAR label.

Estimated Monthly Cost Savings Table for Climate Zone 13:																		
Number of Stories:				Single Story										Double Story				
Foundation Type:		Slab-on-grade			Basement			Crawlspace			Slab-on-grade			Basement			Crawlspace	
Home Size (SF):	1,000	2,000	2,500	1,000	2,000	2,500	1,000	2,000	2,500	2,000	4,000	5,000	2,000	4,000	5,000	2,000	4,000	5,000
Estimated Monthly Savings:	$20	$35	$40	$20	$25	$30	$25	$40	$45	$35	$70	$80	$30	$50	$70	$40	$70	$85

(a)

Builder Option Packages for ENERGY STAR® Labeled Homes[1]

Builder Name:_____

House Address:_____ City:_____ State:_____

Climate Zone 13 [2]

BOP Selected	BOP Number	Window Requirements			Minimum Insulation Requirements[3]						Minimum Equipment Requirements[4]							
							Floor Above				Gas Furnace Htg / Elec Clg		Electric Htg / Electric Clg		Oil Hydronic Htg / Elec Clg		Gas Hydronic Htg / Elec Clg	
		Maximum Window Area[5]	Window U-value	Window SHGC[6]	Attic	Exterior Wall[7]	Unheated Space	Basement Wall	Slab	Crawlspace Wall	Heat (AFUE)	Cool (SEER)	Heat (HSPF)	Cool (SEER)	Heat (AFUE)	Cool (SEER)	Heat (AFUE)	Cool (SEER)
	1	12%	</= 0.35	</= 0.50	R- 38	R- 15	R- 19	R- 10	R- 6	R- 10	90%	10	--	--	82%	10	90%	10
	2	12%	</= 0.40	</= 0.40	R- 38	R- 13	R- 19	R- 10	R- 10	R- 10	90%	10	--	--	82%	10	90%	10
	3	15%	</= 0.35	</= 0.40	R- 38	R- 15	R- 19	R- 10	R- 10	R- 10	90%	10	--	--	82%	10	90%	10
	4	15%	</= 0.35	</= 0.50	R- 38	R- 19	R- 19	R- 10	R- 8	R- 10	90%	10	--	--	82%	10	90%	10
	5	18%	</= 0.35	</= 0.35	R- 38	R- 19	R- 19	R- 10	R- 8	R- 10	90%	10	--	--	82%	10	90%	10
	6	21%	</= 0.35	</= 0.35	R- 38	6.5" SIP	R- 19	R- 10	R- 8	R- 10	90%	10	--	--	82%	10	90%	10
	7	21%	</= 0.35	</= 0.35	R- 38	R-12 ICF	R- 19	R- 10	R- 10	R- 10	90%	12	--	--	82%	10	90%	12
	8	12%	</= 0.45	</= 0.50	R- 38	R- 19	R- 19	R- 10	R- 8	R- 10	94%	10	--	--	84%	10	90%	10
	9	15%	</= 0.34	</= 0.37	R- 30	R- 19	R- 19	R- 10	R- 8	R- 10	90%	10	--	--	--	--	--	--
	10	15%	</= 0.34	</= 0.37	R- 38	R- 17	R- 19	R- 10	R- 8	R- 10	90%	10	--	--	--	--	--	--
	11	15%	</= 0.35	</= 0.40	R- 38	R- 15	R- 19	R- 10	R- 8	R- 10	94%	10	--	--	84%	10	90%	10
	12	18%	</= 0.45	</= 0.50	R- 38	R- 13	R- 19	R- 10	R- 6	R- 10	--	--	2.8 COP	13 EER	86%	10	--	--
	13	21%	</= 0.35	</= 0.35	R- 38	R- 19	R- 19	R- 10	R- 8	R- 10	94%	11	--	--	84%	10	90%	11
	14	21%	</= 0.45	</= 0.45	R- 38	R- 13	R- 19	R- 10	R- 6	R- 10	--	--	2.8 COP	13 EER	88%	10	--	--

BOP Provider Company's Name: _____ BOP Provider's Address: _____
BOP Provider Phone number:_____ _____
BOP Inspector's Name: _____ BOP Inspection Company's Name: _____
Inspection Date:_____ Estimated Monthly Cost Savings:[12]_____

(b)

Figure 12-2 Energy Star description of builder option packages (a–c).

Additional Requirements for Climate Zone 13								
Envelope		Equipment					Design Limitations	
Infiltration[8]	Door	Thermostat[9]	Water Heater Energy Factor[10]	Duct Leakage[11]	Duct Insulation[12]	Ventilation	Above Grade Area per Floor	Window Orientation
<= 0.35 ac/h; blower door tested	>/= R-5	Programmable	>/= 0.56 gas; >/= 0.86 elec;	<= 6% leakage (CFM/CFM) to unconditioned spaces at 25 Pascals; field verified	Insulate ducts in unconditioned spaces to R-6	Active ventilation recommended	<= 2500 S.F.	<= 62.5% of allowable Maximum Window Area (see pg.2) can be located on the south and west

Footnotes:

1) Meeting all the requirements in a Builder Option Package (BOP) qualifies an individual home as ENERGY STAR compliant. ENERGY STAR labeled homes are designed to use at least 30% less energy than the Home Energy Rating System (HERS) Reference Home in the areas of heating, cooling, and domestic water heating. Homes that do not meet the requirements in the BOPs, should be certified by a local HERS rater. Homes built to BOP specifications must be verified by a RESNET-approved BOP provider, in accordance with the EPA/RESNET Agreement on BOPs (see www.natresnet.org/bop/agreement.htm). Additional efficiency and savings can be achieved by selecting other ENERGY STAR labeled products throughout the house (e.g., lighting, appliances). For more information, visit www.energystar.gov. Regardless of these specifications, all local codes must be followed.

2) To determine the appropriate climate zone for the building site, see the 2000 International Energy Conservation Code, Figures 302.1 (1-50).

3) Thermal requirements vary with local building codes. Ensure that insulation levels meet all relevant codes. The BOPs were developed for homes using wood framing, unless otherwise noted [i.e., insulated concrete form (ICF) or structural insulated panel (SIP)]. If metal framing is used, consult a local HERS rater to determine additional upgrades necessary to achieve similar thermal performance, such as additional insulated sheathing.
The insulation R-Value of each component (i.e., attic, exterior wall, etc.) must meet or exceed the required level designated in the BOP. The overall R-Value for components with multiple insulating levels can be determined by calculating a weighted average of the R-Values (based on the percentage of the total area each constituent covers). For example, if the attic insulation required is R-38, and 25% of the ceiling is cathedral insulated to R-19, the required R-Value for the remaining roof would be: 0.75 / [(1 / 38) - (0.25 / 19)] = 57, or R-57. Likewise, if a skylight is used as part of the roof, a similar calculation would be made using the appropriate R-values.

4) Install properly sized HVAC equipment. Recommended sizing methods: size heating & cooling equipment to ACCA Manual S specifications; size ducts to Manual D specifications, both based on Manual J load calculations. Geothermal heat pump equipment is specified in the table by a heating COP and a cooling EER.

5) Maximum window area is a ratio of total window unit area to total above-grade conditioned floor area (WFA). For example, a house with total above-grade conditioned floor area of 2,000 square feet and total window area of 400 square feet has a WFA of 400/2,000 = 20%. Regardless of the maximum window area, up to 0.5% WFA may be used for windows with decorative glass (e.g., doesn't meet U-value or SHGC requirements). Likewise, a maximum of 1.0% WFA may be used for skylights. For example, a house with total above-grade conditioned floor area of 2,000 square feet may have only 10 square feet (0.5% of 2,000) of decorative glass and 20 square feet (1% of 2,000) of skylight area. All decorative glass and skylight window area counts towards the maximum window area designated in the BOPs.

6) Solar window screens may be used to meet SHGC requirements. The overall SHGC for a window unit with solar screen is determined by the following equation: [(window SHGC) x (solar screen SHGC) x (percent of area covered)] + [window SHGC x percent of area not covered]. For example, a window with a SHGC of 0.5, using a solar screen that provides 70% shading (the equivalent of 0.3 solar heat gain coefficient) and covers 60% of the window has an overall solar heat gain coefficient of [0.5 x 0.3 x 0.6] + [0.5 x 0.4] = 0.09 + 0.20 = 0.29.

7) Insulated Concrete Form (ICF) walls must include a minimum 4" concrete thickness with minimum total form insulation of R-12. An ICF wall can be substituted for all BOPs with wall insulation levels <= R-17.
A 4.5" Structural Insulated Panel (SIP) must have an overall insulation level >/= R-15.5. A 4.5" SIP wall can be substituted for all BOPs with wall insulation levels <= R-17.

8) ASHRAE Standard 62-89 requires 0.35 ac/h of outdoor air (but not less than 15 CFM per person) to meet ventilation requirements for residential dwellings. It allows for infiltration and natural ventilation to satisfy this requirement. However, without active ventilation the actual infiltration rate could vary significantly throughout the year. To ensure consistent indoor air quality, it is recommended that homes are built to 0.20 ac/h or tighter and an active ventilation system is installed to achieve a minimum of 0.35 ac/h. To maximize savings, use a heat recovery ventilation system in cold and moderate climates, or energy recovery ventilation in hot climates.

9) Programmable thermostats used in homes with heat pumps must have "ramp-up" technology to prevent the excessive use of electric back-up heating.

10) For BOPs with Oil or Gas Hydronic equipment, domestic water heating must be provided by the space heating boiler (tankless).

11) Duct leakage is determined by: duct leakage (%) = measured leakage from portion of duct system in unconditioned space / design airflow. For example, duct leakage for a forced air system with a design airflow of 2,000 cubic feet/minute and a measured leakage to unconditioned space of 100 cubic feet/minute (CFM) is equal to 100 CFM / 2,000 CFM = 0.05, or 5%. Duct leakage tests such as the blower door subtraction method or simultaneous duct blaster and blower door testing can be used to measure duct leakage to unconditioned

12) A minimum of R-4 duct insulation is recommended for ducts in conditioned space to prevent condensation.

13) See that attached "Monthly Utility Savings" sheet to determine estimated monthly utility savings.

Notes:

a) The symbol " – " means that the option is not available for that specific BOP.

(c)

Figure 12-2 (*Continued*)

ICF walls are specifically listed in the BOPs. Package no. 5 includes ICF walls along with the other things required in almost all the BOPs (efficient windows, good attic and foundation insulation, and so on). In fact, a separate rule in the BOPs rules allows the builder to substitute ICFs for R-17 frame walls in *any* of the other packages.

There is also a money-saving alternative to full testing that is offered to large builders. Builders who submit 75 or more houses for Energy Star certification in a single year can be tested according to a **sampling procedure**. The houses all get inspected visually, but only 15 percent of them are actually tested. If all the tested houses pass, the builder gets certification for all the houses. Of course, the rating fees are much less because there is less work.

Like anything, once you do a few Energy Star houses you get efficient at the process. If you are going to sell to buyers who put a lot of stock in energy efficiency, this can be a great marketing tool. You can list your houses as Energy Star houses, you can use the logo, and you can tell people that they have a chance to get an energy efficient mortgage. You can even get listed in directories as an Energy Star Builder. It pays to look into the program, and consider signing up.

R-2000

Before the United States had Energy Star, Canada had a similar program called **R-2000**. The R-2000 program is a partnership between the Canadian Home Builders Association (CHBA) and the government organization Natural Resources Canada. Compared with Energy Star, R-2000 involves even more testing and training. In addition, it has requirements not only for energy efficiency, but also for indoor air quality (achieved through careful ventilation and use of environmentally responsible materials).

R-2000 awards new houses the title of *R-2000 Certified Home* and the builder right to use the R-2000 name in marketing, if the house meet specific guidelines:

- The builder has completed a special R-2000 training course before becoming eligible for certification.
- The ventilation system installer has also received special R-2000 training and is licensed.
- The house goes through plan evaluation, on-sight inspection, and air tightness testing, all by people who have received special training and are licensed.
- The inspections and tests verify that the house will consume significantly less energy than a house constructed according to minimum standards, and used materials that contribute to good indoor air quality and use of environmentally responsible materials.

The insulation value of the foam in ICFs, the thermal mass of the concrete, and the near-perfest air barrier that ICFs create makes it a lot easier to meet the stringent R200 requirements. Partly for this reason, many Canadian ICF builders find it worthwhile to go ahead and get their homes R2000 Certified.

The sign-up and certification process is similar to that for Energy Star. More information is available through *www.r2000.org*, the web site that Natural Resources Canada maintains on the program. Information is also available at *www.chba.ca/r2000*. This is the section of the site of the Canadian Home Builders' Association that covers the R2000 program.

Energy Efficient Mortgages

Energy efficient mortgages can give borrowers big benefits. One official stated that the buyer of a typical $100,000 house could qualify for an extra $10,000 to $13,000 on the mortgage if the house were certified energy efficient. Depending on the mortgage, the borrower might also get a lower interest rate, a lower required down payment, or reduced closing costs. Table 12-1 presents a specific example based on an energy-efficient house with the following characteristics:

TABLE 12-1   Example of an Energy Efficient Mortgage Provided by the Federal Mortgage Acceptance Corporation

Underwriting factor	Conventional mortgage	Energy mortgage
Purchase price	$100,000.00	$103,300.00
Down payment	$10,000.00	$3,079.00
Mortgage requested	$90,000.00	$100,200.00
Monthly PI qualified for	$660.39	$735.39
Taxes & insurance	$139.61	$139.61
PITI	$800.00	$875.00
Less monthly savings	0	$75.00
Adjusted housing expense-to-income ratio	27.6%	30.2%
Adjusted LTV	90%	97%
Consumer purchasing power	$90,000.00	$100,221.00*

*With the same income, a consumer can afford an additional $10,200 in purchase with energy upgrades while capturing $75 a month in energy savings.

Interest rate	8%
Borrower's monthly income	$2,900
Added cost of the energy efficiency features	$3,300
Projected monthly energy savings from home energy rating	$75

As a builder of ICF homes, the energy efficient mortgages can give you a big benefit, too. If you can show buyers that your house qualifies, they will see that they can save on borrowing costs. Some people who would not otherwise have been able to afford your house will now be able to because they can borrow more money.

But the biggest obstacle to getting this benefit, believe it or not, is *finding* a lender that offers energy efficient mortgages. Steve, the former owner of a HERS company, explains:

> In some areas you can call every local lender or lender's representative and not one of them will tell you they have a stretch mortgage. They're not required to offer them, and so not all of them do, and that's okay. But really shocking to most people is that at a lot of them that technically do offer it—they have it on the books—but the representatives don't know they have it. What happened is that headquarters puts a stretch mortgage in the system. It's there in the database, with specific rates and terms and forms and everything. But they don't include it in the training of their reps because they don't sell a lot of them. So the reps don't know it's even there and they wouldn't know the procedures for preparing the mortgage if they did. So you ask, they say they don't have it, and that's the end of it.

This is important because you, the builder, really need to confirm that there are some lenders who will give your buyers a better mortgage before you start making claims to people. Ideally you want to find some lenders who will give your house a preferred mortgage, give their names and numbers to potential buyers, and when the buyers call, the loan officers understand what the buyers are talking about and are eager to serve them. If instead you say that someone who buys your house can get an energy efficient mortgage, and then buyers call some random lenders who all say they've never heard of such a thing, you are in an awkward situation.

So it makes most sense for you to call around first and find a suitable lender or two. Fortunately, there are a few ways you can try to track them down.

One is to talk to your ICF supplier. Some suppliers saw the problem of finding the qualified lenders and have done something about it. They have established relationships with specific lenders who have agreed to give better terms to houses built with their ICFs. So your supplier may be able to give you the name of a good lender or two.

Another option is to check the Energy Star web site (*www.energystar.gov*). Energy Star has a listing of lenders providing energy efficient mortgages just as it has a listing of builders constructing energy efficient houses. The lenders get listed if they offer a mortgage with better terms for an Energy Star home. Since they've taken the trouble to join the program and get listed, people at these lending companies should know what you're talking about and be able to provide buyers with a loan.

Another option is to ask federal government agencies that provide services to lenders. They support lenders who offer energy efficient mortgages, and they keep lists of them that they share with the public.

One of these is the Federal Housing Administration (FHA). FHA insures many mortgage lenders. If someone defaults on a mortgage, FHA covers the loss. But of course FHA will only do this if the mortgage has sensible terms in it. Well, FHA will insure energy efficient mortgages and it has rules for the ones it will insure. This is referred to as the FHA Energy Mortgage Program. Technically these aren't energy EEMs, but energy improvement mortgages, but they work about the same from the borrower's point of view. The FHA is run by the U.S. Department of Housing and Urban Development (HUD), and your local HUD official is supposed to have the names of lenders who offer FHA-insured energy improvement mortgages in the local area.

Other government sources to check with are the Federal National Mortgage Association (sometimes called Fannie Mae) and the Federal Mortgage Acceptance Corporation (Freddie Mac) (see Fig. 12-3). These are large corporations set up by the federal government to buy mortgages from the lenders that make them in the first place. A big chunk of all the mortgages that lenders in the United States provide to the public they just turn around and sell to Fannie Mae or Freddie Mac. Well, these two corporations will buy energy efficient mortgages and they have specific rules for the ones they accept. They also keep listings of which of their lenders offer those mortgages.

Fannie Mae provides its list on its web site, *www.efanniemae.com*. The site has a page entitled "Mortgage Broker Center" that has a link to check out "Products." EEMs are one of the products. The page on EEMs includes a link to a list of lenders, as well as contact information for HUD officials to talk to about the program.

Energy Efficient Mortgage Comparison

PROGRAM NAME	OLD ENERGY EFFICIENT MORTGAGE	MY COMMUNITY EEM	EEM
COMMITMENT VOLUME	None specified, underwriting adjustment	Product in MCM Suite	Full-fledged product
PROGRAM TERM	None	None	None
ELIGIBLE LOANS	15 or 30 year Adjustable or fixed-rate Purchase or Refinance	Same	Same
ELIGIBLE PROPERTY TYPES	1-4 unit Owner-occupied New construction or existing homes	One-unit	One-unit
FIRST MORTGAGE MAXIMUM LTV	95%	100%	100%
MAXIMUM CLTV	95%	105%	105%
TOTAL DOWN PAYMENT	5%		
MIMIMUM BORROWER CONTRIBUTION	5%	Lesser of 1% or $500 for borrowers at 100% AMI or no income limit in FannieNeighbors areas	3%
ADDITIONAL DOWN PAYMENT AND/OR CLOSING COSTS		Secondary financing must meet Fannie Mae's standard Community Seconds Program guidelines	
VALUE OF ENERGY EFFICIENCY MEASURES, IMPACT ON LTV	For new homes or efficient "as is," the present value of the energy efficiency measures is added to the lesser of purchase price or appraised value. For retrofitted homes, the installed cost is added to the lesser of purchase price or appraised value The LTV calculation is based on the lesser of the adjusted value or the adjusted purchase price.	For new homes or efficient "as is," the present value of the energy efficiency measures is added to the appraised value. The LTV calculation is based on the lower of purchase price or adjusted value. For retrofitted homes, the installed cost is added to the purchase price and the appraised value. The LTV calculation will use the lower of the adjusted purchase price or the adjusted value.	For new homes or efficient "as is," present value of energy efficiency measures is added to the appraised value. The LTV calculation is based on the lower of purchase price or adjusted value. For retrofitted homes, the installed cost is added to the purchase price and the appraised value. The LTV calculation will use the lower of the adjusted purchase price or the adjusted value.

Figure 12-3 Comparison of different generations of energy efficient mortgages from the Federal National Mortgage Association.

FannieMae

Energy Efficient Mortgage Comparison

ENERGY SAVINGS PRESENT VALUE	Calculated using the monthly equivalent of the mortgage rate for fixed rate mortgages or the fully indexed accrual rate for adjustable mortgages for a term not to exceed the weighted physical life (in months) of the Energy Improvements.	Calculated by the HERS rater using a set mortgage rate, determined annually by Fannie Mae, for a term not to exceed the weighted physical life (in months) of the energy improvements.	Calculated by the HERS rater using a set mortgage rate, determined annually by Fannie Mae, for a term not to exceed the weighted physical life (in months) of the energy improvements.
DU SOLUTION	N/A	DU Approve or My Community 100 Plus Guidelines Energy savings are added to income in the "Other income" section. For new homes or energy efficient as is, add present value to the appraised value of the home. DU automatically uses the lower of the purchase price or adjusted appraised value. For retrofit homes, add installed cost to "alterations" section and it will be added to the purchase price. Also add the cost to the appraised value. DU will automatically take the lower of appraised value or purchase price.	DU available Energy savings are added to income in the "Other income" section. For new homes or energy efficient as is, add present value to the appraised value of the home. DU automatically uses the lower of the purchase price or adjusted appraised value. For retrofit homes, add the installed cost to the "alterations" section and it will be added to the purchase price. Also add the cost to the appraised value. DU will automatically take the lower of appraised value or purchase price.
MAXIMUM ALLOWABLE DEBT-TO-INCOME RATIOS	Depends on underlying product	41% single ratio N/A when DU Solution used	41% single ratio
ADDITIONAL INCOME SOURCES	None	Manufacturer appliance rebates can be used toward closing costs. Tax credits, where available, can be used toward borrower's income.	Manufacturer appliance rebates can be used toward closing costs. Tax credits, where available, can be used toward borrower's income.

CONTACT: Michelle Desiderio, Fannie Mae Senior Product Developer at michelle_desiderio@fanniemae.com or 202-752-4041.

Figure 12-3   (*Continued*)

Other Energy Efficiency Award Programs

Some states and provinces also have programs to reward energy-efficient construction, and from time to time the federal government passes new ones. It is hard to generalize about these because they vary widely and may change without warning. But it is a good idea to keep your eye out for them and see if your buildings might qualify. This will take some legwork for the pioneers, but it can pay off in the longer term.

Scott, a consultant in Oregon, found some lucrative programs that have paid back well for ICF buildings in his area.

> One thing I found early on was that people would point to the incentives but not many would do anything about it. Oregon has a progressive energy efficiency policy and they have a tax credit system that's mostly used for HVAC upgrades, irrigation pump projects, and so on. Basically, you get a tax credit of 35% of your cost for the extra cost of energy-efficient equipment you build into a project. I was able to demonstrate and receive the first one of these tax credits for ICF walls. I had to go through the full energy calcs to prove that ICFs were energy efficient enough to meet the program requirements. It took a year and a half. Now I do a steady stream of these projects, where I do the legwork for the application, get my clients a big rebate, and they save a lot on their fuel costs, too.

Leadership in Energy and Environmental Design

A nonprofit organization called the U.S. Green Building Council (USGBC) has come up with a system to allow people to rate and compare buildings according to how environment-friendly their construction materials and methods are. It goes by the name of **Leadership in Energy and Environmental Design** (LEED). It is designed to encourage green building by giving a marketing advantage to structures that meet high environmental standards. As with Energy Star, showing off a LEED certification can attract interest and buyers.

A good example of a LEED project using ICFs comes from Jim, a general contractor in Pennsylvania:

> It was a building for the Pennsylvania Department for Environmental Protection. We were hired based on our ability to get LEED certification. It's one of their requirements. We went with ICF walls because they're cost-effective and the R-value is high and they got us fast to a waterproof building that's ready for drywall. The R-value helps toward getting the LEED rating. The other things are useful to the contractor.
>
> LEED is concerned with energy use—that and total resource use are primary factors in getting your rating. Whether the materials are locally manufactured is also weighed in. If materials are manufactured within 500 miles of the job site, that counts toward the rating. Our forms were made in Pittsburgh, so that helped, too.
>
> The main difference in a LEED building is that there's a lot of paperwork—much more than usual. There are guidelines for the LEED award at the U.S. Green Building Council. They have a really good breakdown of how it works. I bought the book they have. It takes a couple of months of study to understand. If I were coming at it again, I'd tell our people to hire someone to take care of the application and record keeping instead of me. Our architects were very up on the program and that

helped, but architects don't always know the costs. They'll go after points that cost $25,000 and not after some others that cost $5,000.

I think my biggest advice for someone else doing a LEED building is to have a design team from the beginning and set the LEED goal at the beginning of the project. You get it based on what you design into the building and how you build. If you decide to go after certification halfway through, it will be very difficult to get.

Showing the LEED certification may be a plus for some owners. There are also now some owners who set out to construct a new LEED building. So they actively seek contractors familiar with construction methods that rate high on the LEED scale. Since ICFs can help on the rating scale, the ICF contractor has an advantage bidding on these projects. Now, there are also some government incentive programs that offer builders or owners financial rewards for constructing a building with LEED certification.

ICFs provide energy efficiency, use local materials (concrete), and may contain recycled materials in the concrete. All of these things count in the LEED rating system. So it is a bit easier for ICF buildings to get LEED certification than for some others to get it.

The LEED program applies to commercial and large structures. To qualify, a building has to fall into one of the standard building code categories for offices, retail and service establishments, institutional buildings (e.g., libraries, schools, museums, churches, etc.), hotels, or residential buildings of four or more habitable stories. The USGBC is now developing programs for existing buildings, interior construction, speculative buildings (shells), homes, and communities.

If you want LEED certification for a building, the first step is to register the project with the USGBC (Go to *www.usgbc.org*). The next is to submit documents on the specifics of the building. A special USGBC committee will use these documents to score the building. The paperwork gets involved, so it pays to find and hire a **LEED accredited professional**. This is someone trained to do the job of gathering all the information and working with the USGBC staff. There are a limited number of them, and the USGBC can supply their names. The submitter is responsible for paying all costs of gathering and providing the information to a committee of the USGBC. The submitter is usually the building contractor or the owner—whoever is applying for the certification.

The LEED rating system works on a scale with a maximum of 69 points (see Fig. 12-4). A building can receive different levels of certification, depending on exactly how many points it gets. The levels are as follows:

Certification	26–32 points
Silver	33–38
Gold	39–51
Platinum	52–69

A building gets evaluated on various attributes that fall under five basic categories: sustainable sites, water efficiency, energy and atmosphere, materials and

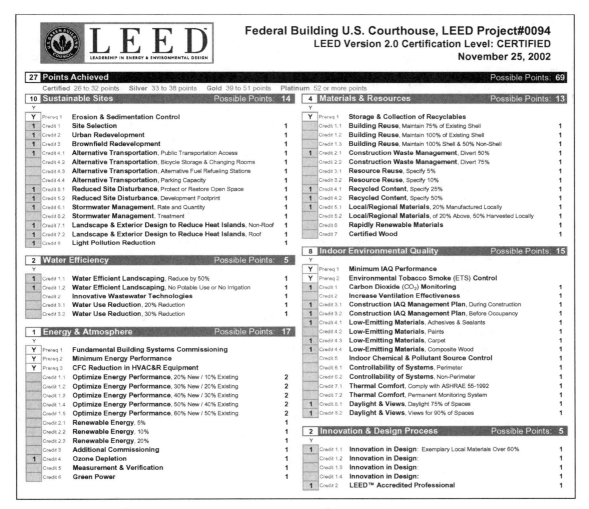

Figure 12-4 Standard LEED certification form. (*Courtesy of U.S. Green Building Council.*)

resources, and indoor environmental quality. The building must have certain minimum attributes in most of these categories, and then points get awarded for additional, nonrequired attributes.

The applicant lists the attributes he/she believes the building fulfills on a form submitted with all the documentation. After all the materials for a building are submitted, the USGBC responds with its judgment of which points it will allow. There are several rounds of appeal, in which the applicant can provide more information and ask for more consideration on some of the scoring. In the end the USGBC committee states a final score and provides the applicant with a plaque and certificate that display it.

Insurance Rates

In most cases it's easy to get a reduced insurance premium for an ICF house, and the savings can be substantial. According to Chris, a homeowner in Ontario:

> I realized an approximate reduction of 25%–30% in my home policy. But I would like to explain that it was not due to the ICF per se, rather it is what's behind the "foam"–concrete. My agent simply looked in his guide and entered the applicable rate. A "stick built" home will cost more for structural insurance than the concrete home. For the concrete home, the inevitable payout from insurers due to fire, flood, storm damage, etc. has been shown to be far less than that for the other.

Home insurance rates are based on how much the insurer has had to pay out for similar houses in the past. If houses of a particular type or in a particular area suffer damage less often, the insurance company has to pay out less money for those houses and it charges less for the insurance.

Concrete houses like those built of ICFs get reductions on two counts. First, concrete homes have historically suffered less damage in fires. So they get a reduction in the premium for the fire portion of the insurance. Second, they suffer less damage from many types of natural disasters (like hurricane), so they get a lower rate for that.

The exact savings vary. They depend on what the rest of the house is like and how common certain types of damage are in the local area. Builders have reported reductions in homeowner policy premiums of anywhere from 15 to 48 percent. According to Henry of Florida:

> I just completed an ICF home in Florida. It's located in the 110 mph coastal area. The house is two story with wood roof trusses and a tile roof. USAA insured the structure for approximately $500/year. For comparison, my builder's risk insurance during construction was $2,100 per 6 months. I should add that the house has impact resistant windows (hurricane resistant) and this lowered the rate a bit also. The other homes in my neighborhood are either concrete block or frame or a combination of the two. I'm guessing but their rates are probably twice what I'm paying.

Some insurance companies now have ICFs listed in the database to come up with insurance rates for them. But the agents might not all be aware of it, so you may have to help them understand the product and figure out how it might be listed on their computer. The companies that do not list ICFs will usually have another category that they apply to concrete construction, and they will usually give you that.

Of course it's not the contractor's job to find out where the homebuyer can get the best insurance rates. But as with energy efficient mortgages, it may still pay you to get the names of some agents that can quote a decent break to the buyer. It helps the buyer go directly to an agent that gives a discount, instead of first happening to talk to one or two who don't. It is also a good marketing tool because it gives the buyer the names of people who provide independent confirmation that yours is truly a superior house.

Working with Building Officials, Engineers, and Architects

Overview

To get any building constructed you need more people to help out than just the contractor and the buyer. The president of one construction materials firm said that to get a new house built with a new product you need the approval of "the ABCs," which includes:

- The architect
- The building official
- The contractor installing the new product
- The developer
- The engineer
- The financier
- The general contractor
- The homeowner

In truth not all of these parties are involved in every new house project. And roles are different for commercial construction. But the point is a good one. To get an ICF building constructed often requires convincing several different people that the product is a good one, that it won't be too hard to work with, that there are no serious risks, and so on. As ICFs become more common, more and more of these people are already familiar with them. But especially when ICFs are new to your area you are going to find that you have to spend some time bringing some of these professionals up to speed and making them comfortable.

In homebuilding the one person whose blessing you almost always need is the local building official. If he says no, it's very tough to get permission to use a new product. If he says okay but he's skeptical, you may have to take time-consuming and costly steps to keep him satisfied. One typical requirement set by building officials is that the plans have to be engineered—analyzed and okayed by a licensed structural engineer. This is getting rarer and rarer, but it still pops up every now and then. And when it does, you will have to bring the engineer up to speed, too.

Most houses are constructed without the services of an architect. But buyers of higher-end homes hire them more frequently, and since many ICF houses are at the high end, you stand a good chance of being asked to work with an architect. If the architect has worked with ICFs before, there's nothing to it. If not, you go through the same education process.

In commercial and larger residential projects, the order is usually reversed. Design decisions are almost always in the hands of an architect. The architect specifies what the building will be constructed out of. An engineer analyzes the building to fill in certain detailed specifications that ensure the building will be structurally sound. By and large, the local building department accepts the sign-off of the architect and engineer. It is rare that the building official challenges any design or choice of materials that the designers have specified. After all, these designers are licensed and they carry the legal liability for things that go wrong with the final building.

Fortunately, all of these people are usually reasonable, and there is now a lot of time-tested advice for working with them.

Building Officials

Except in a few areas, the general contractor building a house has to work with the local government building department so that the officials there can examine the plans and inspect the actual construction. They do this to verify that it meets their basic requirements of safety and quality, and stop work when it doesn't. When you submit plans that include products they're not familiar with, they can accept, reject, or require more information.

Your first course of action should be to look for a building official who is familiar with ICFs. That person is already educated so you don't have to do it. So try to find out who on the department staff has seen them before. Then try to submit your plans to that official. If that is not an option, at least ask the experienced official to advise the inspector assigned to your project.

This is not hard to do. Go to the building department offices as early as you can—even when building with ICFs is just an idea. Show the staff some of the literature about ICFs and ask for their thoughts on the product. Don't tell them about it—*ask*. If someone has seen them before, you'll hear about it. And if you keep the conversation going you'll also get some very valuable information. You'll learn what concerns they might have about ICFs so you can come back the next time prepared.

If no one is familiar with ICFs, the second part of the conversation is all the more important. Ask what they would need to see to approve a project with the product. Then you know what you'll have to do when you bring in an actual set of plans.

Duane, a project manager with a homebuilder in Texas, gives his experience working with the local building departments:

> We had to have an engineer's stamp on our plans originally. But now every building department in Texas has seen this and approves it. We have to give them documentation and work with them a while, but now we don't need a stamp on the plans.

Lambert, an ICF contractor in Minnesota, has a similar story:

> Normally all the building department requires is a blueprint. If they don't trust it, they just ask for an engineer's stamp. I only needed a stamp a couple of times. Usually the user's manual, model code evaluation reports, and good answers to a few questions are enough.

Generally speaking, the biggest requirement a building official will lay down is that the house has to be engineered (or as some put it, the plans have to have an "engineer's stamp"). This is also the requirement most contractors try hardest to avoid, since the engineering takes time and costs money—from a few hundred to a few thousand dollars. They ask for engineering because of the peace of mind it gives—a trained, licensed professional has looked at the product carefully, checked the plans, and made the necessary changes to produce a sound building, and has signed off and taken legal responsibility. In fact, in some cities nowadays *every* house has to be engineered, so there is no way out of this.

But in most places you only get asked to provide engineering if the building department has questions about the project which they don't feel qualified to answer themselves. And there are often less expensive ways to reassure them than a full engineering analysis. They may ask for one of these simpler measures. And even if they ask for engineering, they might change their minds when they see all the information you can provide.

For starters, a good set of plans often gets the building official favorably disposed to your project. Include details of key parts of the ICF wall. Many times your ICF supplier can provide you with standard details. There are also details in the ICF section of the *International Residential Code*. The International Residential Code is the major model code that most local codes are expected to come from in the future.

One important piece of information you can give the official is the **evaluation report**. This is also called a **research report**. It is a sort of assurance that a new product meets the building code.

The local building code—the rules of construction that the inspector is working on—is almost always based on one of the major *model codes*. In the United States there are four model codes that are widely used, called the **major model codes**. They are the following:

- Basic Building Code (BBC)
  - Produced by the Building Officials Congress of America (BOCA)
  - Used mostly in the Northeast

- Standard Building Code (SBC)
  - Produced by the Southern Building Code Conference International (SBCCI)
  - Used mostly in the Southeast
- Uniform Building Code (UBC)
  - Produced by the International Congress of Building Officials (ICBO)
  - Used mostly in the West
- International Residential Code (IRC)
  - Produced by the International Code Council (ICC)
  - Designed to replace the other three codes everywhere, and increasingly adopted around the United States.

When a product is so new that it is not in the building codes, the sellers are permitted to submit test results and data to show that it works in a building in a way consistent with one or more of the model building codes. If everything is in order, the model code organization publishes an evaluation report on the product. The evaluation report is usually three to eight pages long. It tells the local building official that the product is acceptable and lays out the steps the official can use to inspect the product and ensure that it is installed correctly.

An evaluation report is often enough to make the local official comfortable with ICFs and accept their use. Typically the building inspector wants to see an evaluation report from the one model code that the local government in the area has adopted. But there are cases of inspectors being satisfied with a report from one of the other model code organizations.

You can get evaluation reports for many brands of ICF now, because many of the manufacturers have gone through the work of getting them issued. Ask your local distributor for some copies. In fact, it's a smart question to ask before deciding which brand of ICF to use. Builders who find out that the local official needs technical information and then discovers that his brand of ICF has no evaluation reports, often wish they had taken this into account when choosing their ICF.

A document that may even be more powerful nowadays than an evaluation report is the model code itself. ICFs are now covered in the newest versions of some of the model codes because the industry (led by the Portland Cement Association, the Insulating Concrete Forms Association, and the U.S. Department of Housing and Urban Development) has produced the necessary technical data and submitted changes to the *International Residential Code* and the *Standard Building Code*. These model codes now include sections on ICFs. In the IRC they're in section R404 (which covers the use of ICFs in below-grade construction) and R611 (which covers the use of ICFs in above-grade construction).

The model codes lay down the requirements for use of ICFs in homebuilding. Much like the evaluation report, they show the building official that ICFs are an accepted product and show him what to look for when looking at the plans and the construction, so he can be confident he is approving a sound structure. But they are more integrated into the rest of the code and carry more weight than an evaluation report.

In Canada the situation is a bit different. There is one national model building code, the National Building Code of Canada. The various provinces of the country adopt this code, making some changes to suit their local situation. Currently the National Building Code of Canada does not include ICFs. However, the next revision of the code is due in 2005, and it is slated to include them. In the meantime, you can present the local building department with a part of the code called *NBC Part 9: Housing and Small Buildings*. This includes rules for constructing buildings with walls of reinforced concrete. Many ICFs fall under these rules because they are simply reinforced concrete walls with foam on each face. So the building official may be satisfied to follow that. Better still, many ICF companies have Canadian evaluation reports that most officials accept. These are issued by the Canadian Construction Materials Centre (CCMC). Check to see whether the brand of ICFs you want to use has one. If not, check the local building departments to see whether they are likely to require one or whether they will accept other documents.

One other way to bring an inspector up to speed comes from Charles, a building official in New Jersey:

> Plans for a house with ICFs hit my desk a couple of years ago and I didn't know them from Adam. But the builder got me in touch with another inspector two towns over who had done some. That was great. The guy knew what we look for and got right to the crux of the matter. Once he told me what to do we approved it and it worked out fine.

So helping your official understand the product may be as easy as asking around for another inspector nearby who's familiar with the product and having the two of them talk.

Engineers

In home projects, engineers usually work for the builder or the owner who comes to them because the local building department wants an engineer's scrutiny on the building. In commercial projects, they are usually working for the architect, who hires them to work out the structural details of the building. In either case, they rarely veto a new product that is well documented because, frankly, it's not really their choice.

And ICFs are now well documented. The time was when engineers did sometimes object strenuously to the use of ICFs and got them taken out of a project. But that was because they didn't have the data to figure out the structural properties of the ICFs reliably. Now the data and the basic engineering calculations are pretty thorough and widely available.

Actually, any flat ICF system has exactly the same structural properties as a conventional reinforced concrete wall that has been formed up with plywood or steel or aluminum floors. That's because, structurally, it is the same. It just has some foam attached to each face. A lot of times you can just point this out to a skeptical engineer and it will turn him around 180° in one minute. It's pretty obvious when you think about it. And they were trained to engineer reinforced concrete structures from college on.

But they may still want to see all the technical backup because they have probably never engineered a concrete *house*. The forces acting on houses are sometimes different and usually lower than those acting on larger buildings. And houses are built to somewhat different standards. So the engineering calculations are different. The basic calculations for the use of ICFs in houses will help them do the analysis for the project without having to reinvent the wheel.

And if your ICF is a waffle or screen type, the calculations are all different. The wall has a different shape from flat, and it has different strengths that must be taken into account in the engineering. In fact, because it is less familiar engineers may have more resistance to using a waffle or screen wall. But the engineering backup is there to help convince them.

One thing they may want to see is the model code. That's available to them from the model code organization and is probably already sitting on their shelves. You just tell them about it and show them the section on ICFs. It has the basics and covers all types of walls.

Another is the engineering study that the ICF sections in the model codes were based on. That is called the *Prescriptive Method for Insulating Concrete Forms in Residential Construction*. It is available from the Portland Cement Association, Insulating Concrete Form Association, and the U.S. Department of Housing and Urban Development. You can order it off any of their web sites. Try to get the Second Edition, since it also covers high-seismic areas.

Engineers might also be interested in the engineering documents of your ICF supplier. Before the *Prescriptive Method* a lot of ICF companies did their own engineering from scratch, and they usually still print the information in one of their manuals. Ask your local distributor for copies. But most people skip this information now and go on to the other documents.

One other very valuable source of information is the technical staff of your ICF supplier. A lot of the ICF companies have an engineer or two working at headquarters who is happy to talk with other engineers. This is especially useful if your local engineer is stuck on some specific issue or has his own vague doubts about the system. The ICF company engineer can usually clear up specific problems fast, and tell the story of ICFs in the engineer's language, which gives him comfort.

When the project is a commercial or large residential building, dealing with the engineer is mostly the same. For certain kinds of commercial projects the formulas and data for ICFs are not as fully worked out. There is less on it in the codes. And fewer ICF companies have evaluation reports that cover commercial projects. So there is a little less support information. But this is not usually a big obstacle. In commercial and larger projects the engineering work is done more from scratch, anyway. Unless your ICF is missing some critical testing or approvals for the particular project, the engineer will probably not have too much trouble dealing with it. When you seem to hit a snag, contact the technical staff at your supplier again and have them talk engineer to engineer.

One thing to watch out for is *overdesign*. This problem crops up mostly in houses. Engineers are not accustomed to using reinforced concrete in such small buildings, so they may specify thicker walls and more rebar than are necessary. According to Will, an ICF distributor in Georgia:

> Engineers that aren't familiar with ICFs, you can give them a set of plans that are all worked out according to the tables of the *Prescriptive Method*, and they'll want to change a lot just to be safe. But you've got to bring these people along. If you pay them to go through all the tables the first time they'll get familiar with the rules. If you keep bringing them work they come up to speed and start to review stuff later much faster. Eventually you bring the plans to them and they glance them over and say "This is fine." By then they've done it so many times they know what's needed off the top of their heads.

Architects

Architects are involved in almost every commercial and large residential project, and less often in single-family homes. But when they are assigned a project, they pretty much call the shots. It can even be hard for the building owner, who is paying the bills, to overrule the architect in some cases. According to Alex, a contractor in Rhode Island:

> Once we received an inquiry by a world famous architect questioning the structural capacity of ICF walls. What was surprising though was that he was asking because he had designed houses in a high-end development with SIPs [structural insulated panels], and he was looking into switching over to ICFS, and he was not sure if ICFs were as strong. It was hard to convey the idea that ICFs are nothing more than a steel reinforced concrete wall with foam insulation. Depending on how you design them you can make them as strong as you want them to be, people even use them for safe rooms in tornado areas. The funny thing was that the owner group was totally convinced of the benefits of ICFs, they were also putting pressure on the architect but he continued to be skeptical. We were even considering hiring an engineering-consulting firm to confirm the structural capabilities of the system.

The architect is not being arrogant. The architect is hired to make professional judgments on the building materials and design, and to some extent to stick behind what he thinks best.

It follows that when you need to explain ICFs to an architect, attitude is important. You don't want to look like you think you know more than he does. If you seem to be implying that he doesn't know his building materials it's like saying he can't do the job he's being paid for. You're not trying to demonstrate that he's incapable, just filling him in on some details of a particular new product that he needs so that he can consider the product in his design.

For the most part, the information the architect needs is the same as the building official and the engineer. So you might give architects some of the same materials. However, it is very important to remember that architects tend to think much more *visually* than any of these other people. On the top of the stack of things you provide, include those glossy brochures from your ICF supplier. Put in some big, crisp photos of other ICF projects, preferably ones that are really stunning. If there are technical documents that use graphs and charts and diagrams, these are better than the ones that are all text and formulas and tables. Talk to the architect about how easy it is to make curved walls and irregular angles with ICFs. When he sees ICFs can do some of the really interesting aesthetic things, the architect is likely to look more favorably on them and be receptive to learning about them.

And don't forget the value of having him speak with one of his own kind. Ask your contacts for another architect who used ICFs and liked them. Offer this person's name up as a useful contact. An architect may explain ICFs to another architect in a much more familiar way than you could.

When you have a choice

Sometimes you're constructing a building on speculation. So you pick the architect and engineer yourself. And sometimes a buyer asks you for names of good people.

When you get to choose the architect or engineer, the first rule is of course to pick someone who has worked with ICFs before and likes them. Experienced people will work faster and often be cheaper because they have less to learn. Ask the usual sources (distributor, other contractors, listings on ICF-related web sites) for these kinds of people operating in your area.

If you can't find someone experienced, try to get someone who at least seems interested in ICFs. As in any field, there are architects and engineers who like learning new things and others who can't be bothered. It's tempting to pick the "famous" designer even if that designer expresses some doubt about the new product. And this person's view might turn around over the course of the project. But if it does not, you'll have a lot of work and aggravation on your hands. You might have to come back repeatedly to argue that ICFs should be left in the project and not be taken out, you might have to produce never ending streams of documents to answer every little question, and you might have to play the role of the "bad guy" in the project who argues every point.

It's better to find someone who's eager; someone who gladly spends the extra time to come up to speed because it's interesting and exciting; someone who seems to catch on to the idea of ICFs quickly. And nowadays there are more and more of them out there.

14

Marketing

Overview

So you've been trained and found a supplier and are all ready to build with ICFs. Now how do you get work?

If you're planning to be an ICF subcontractor you probably are getting into this because there are ICF buildings going up and you want to get some of that work. So you let all the general contractors in the area know that you're available and qualified. And you stay in touch with them and high on their radar screens any way you can, just as you did when you were a framing carpenter or plasterer or flatwork contractor.

If you are a general contractor, you are probably adopting ICFs because you want to get an edge over your competitors who still use frame or other building systems. You want to make it clear to the public that you can produce a better building for them. For a general contractor, the options for promotions are a lot more varied. There's a lot you can do to attract the public and architects to you.

But don't just run off and start talking to people. If you want a big bang for your buck—that is, lots of serious leads without spending too much of your time—it pays to sit down and think a bit about the main questions:

- Who am I trying to attract?
- What message do I want to deliver?
- How do I deliver it?
- Where do I deliver it?

Life's too short and promotion is too expensive to try everything and then find out that most of it was useless. Pick your targets and you can home in on them precisely and efficiently.

Target

For starters, are you going after residential or commercial construction? The two are pretty different, so odds are that you aren't going to be doing both initially.

If it's houses you're after, you need to get your message out to the home buying public—adults, mostly married, in their late 20s and older. If you want to build larger and commercial structures, you will be courting mostly architects.

Try to pin it down more. If you'll be doing houses, do you intend to go after higher-end houses? That's the segment that has traditionally been larger for ICFs, but some of the others are up-and-coming. Regardless, get as specific as you can so that you can go more directly to your class of buyers.

If you're after large and commercial buildings, is there some type or group of architects that specializes in the kinds of buildings that are likely to use ICFs?

Once you've got your target in mind, you are ready to think through the *what*, *how*, and *where* parts of your promotion. You just turn those questions into the following:

- What is my target interested in?
- What sort of presentation or delivery do they find convincing?
- In what forum or situation can I get them to pay attention to me?

The Message

Even though you may end up giving long presentations, you need to have a short, focused message in mind when you do. You may give the customers a lot of information, but you want those people to come away with a couple of clear conclusions that will stick with them when all the other facts and numbers have become hazy in their memory.

High-end homebuyers most often cite energy efficiency and disaster resistance as the reasons they buy ICFs. In coastal and tornado-prone areas the disaster motive becomes stronger. In other locations energy is usually cited first. So those are usually the things you want to put in front of the customers' minds when they think of ICFs. Mind you, in your area you may find that something else is a higher priority. Don't ignore that. If your people want to talk about quiet, tell them about quiet. Whatever they are most interested in, make that your main message.

But this can be difficult. There are a lot of benefits from ICFs. Do we just ignore the ones of secondary interest? Do we pretend they don't exist? Doesn't the customer want to see that there is a long list of reasons to favor ICFs?

One of the better ways to handle this seems to be to mention everything, but with clear emphasis on the one or two top items that are most important to your customers. A lot of very effective promoters repeat one cover phrase, like "builds a better house," over and over. But when they go to explaining what's better about it they have one or two advantages that they spend sixty percent of their time citing, explaining, and repeating. They go through the other items much more quickly.

The same general rules go for affordable homes and commercial construction. When ICFs sell to affordable housing, it's usually part of a housing program where the energy and insurance savings of the house offset the higher cost of initial construction. Now that makes for a clear message: "a better house for lower payments." Use your own words, but get across the key idea of importance to the buyer. Under a general heading like that you can go into the details of the savings at length, and also mention the many other benefits that are icing on the cake.

In commercial construction the advantages most typically cited are that ICFs provide high strength and energy efficiency less expensively than other systems do. One company started using the phrase "high-performance wall systems" when they pitched to commercial buyers. That might lead into a detailed discussion of energy efficiency and strength, with the rest of the benefits listed as side benefits. But again, tailor the message to what's most important to the audience. If you're going after, say, movie theaters, you will probably include sound reduction in the main message.

The Medium

There are a lot of materials and formats you can use to deliver your message. But remember two basic principles:

- For homeowners, keep it simple.
- For architects, keep it visual.

Homeowners have some interest in the technology, but don't tax them. They may not want to become as big an expert as you do. Architects, as we've said before, respond well to crisp photos and artistically designed graphics.

Somewhere in history it seems that someone decided that every presentation had to have lots of slides with nine bullet points each and lots of tiny type, and every brochure had to be 10 percent picture and 90 percent words. But the truth is that, those kinds of things make most of us go to sleep. The splashy photo, the color graph, and the short slogan do a lot more to spark someone's interest and help them absorb your message. That's true for both homebuyers and architects. The thing to remember about architects is that they expect the visuals to be even more professional looking, and to be backed up by the big data book (which they may never actually read).

Fortunately, there are plenty of materials available that are all ready for you to use. Many ICF manufacturers have very slick brochures that you can get stacks of. Some of these have a place to put your company name. Or you could staple on a card. Ask your distributor what your company offers.

The Portland Cement Association sells pamphlets and attractive two-page Tech Briefs that lay out the advantages of ICFs. You can order small or large quantities of these on PCA's *www.concretehomes.com* web site (see Fig. 14-1).

PCA also sells something called the Concrete Homes Consumer Advertising Toolbox. This is really for organizations that want to mount paid promotional

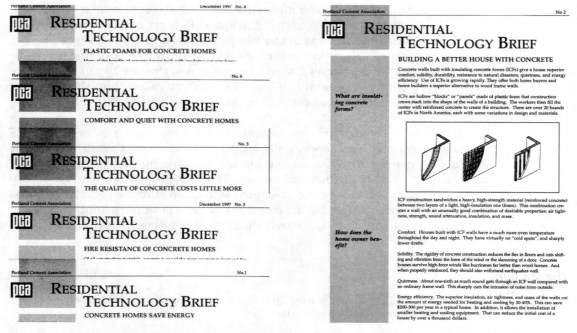

Figure 14-1 Tech Briefs from the Portland Cement Association lay out the advantages of ICF homes. (*Courtesy of Portland Cement Association.*)

campaigns in their local market area. But there are plenty of things you can lift out to make your own flyers or include in your own presentations. The kit includes a CD-ROM with print ads and their components; a cassette tape and digital audiotape with radio ads; and a videotape with footage for television ads.

Home shows

One place to shine is the local home show. First think through whether your target customers will be visiting this show. If they are, it can be a great opportunity.

Make sure you have something visual to attract them. That could be a set of forms stacked up to make a mini-wall or a huge enlarged photo of an impressive project. Then have your materials to hand out and your message ready to deliver to everyone who is interested.

Do home shows only if you're in the business for the long haul. Very few shoppers at a home show will see ICFs and decide right away that they want a house built of them. More likely, you will get calls 1, 2, or 3 years later from people who have finally gotten around to building their "dream house," and now are ready to talk.

Presentations

Sometimes builders find they can interest the local homebuilders' association or the garden club or the local chapter of the American Institute of Architects or other gatherings in getting presentations on ICFs. ICFs are new and different and exciting, so they're a good topic for some of these organizations to put on their program.

If you can arrange one of these with a group that fits your target profile, it can be a great opportunity. You can show a few photos, discuss what's so great about the product, and explain your company and how you can help them. It's hard to get a better opportunity to deliver a pitch.

But if you do one of these, *be prepared*. As good as it is to deliver an impressive presentation to a group, it is equally bad to stand in front of them and be disorganized, confusing, or sloppy. The general rules to a good presentation are as follows:

- Know your stuff. If you can say it in an organized way without hesitation, they can believe it.

- Be enthusiastic. If you are, they will be, too.

- Interact with your audience. If you walk among them, shake hands, and so on, they stay alert and see you are approachable.

- Highlight your company and leave contact information clearly displayed.

- If you have something for them to look at, make sure everyone can see it.

- Even if you're a natural speaker, practice the presentation in advance until you get it down. Try to do at least one full run-through in front of live people.

- Don't get into arguments. If someone disagrees strongly with you or wants to debate some fine point, offer to talk more about it later. If you debate on the spot you'll just turn the rest of the audience off.

Press

Advertising is too costly for most contractors, but there are some ways to get free press coverage. If ICFs are new to your area they are also probably newsworthy. On one of your first projects it's worth calling up the local newspaper and TV reporters and invite them visit the job site and do a story. You can often interest them.

But you have to be a little clever about it. Describe what you're doing in a few words that make it exciting. Keith, a general contractor in Massachusetts, said:

> I figured the reporters wouldn't understand this stuff so easily, so I thought of what else is made of foam. Well, coffee cups are. So I called the paper and left a message for the real estate editor that we were building a house out of concrete and coffee cups. She was intrigued, and we invited her down to the site. She came and wrote the story.

If you succeed in getting someone to cover your project, make sure you give your company name and contact information. Make it clear and repeat it. You want them to print enough information in the article so that people who are interested can contact you.

Parade of homes

One of the more successful marketing tools in residential construction is to build an ICF house for a local Parade of Homes. You can build a beautiful home, hundreds or thousands of people will see it up close, and you can have plenty of displays and handouts on hand to make them appreciate that it's a better house and you know how to build them that way.

But beware. Building in a Parade of Homes house is a major undertaking. It is expensive, time-consuming, and you may never get all of your money back. The crowds are often big, but very rarely will they ask you for an ICF house right away. The advantages are more long-term. The people who toured will realize that you know what you are doing. And when they finally get around to building their next house, they are that much more likely to call you.

Riding the Wave

You can get the benefits of the big promotional programs without the cost. In many areas there are groups like a concrete promotional organization that arrange shows and displays and panels to inform the public about ICF construction. They want to increase awareness and sales of ICFs. But if you get involved it might also increase awareness and sales of your business.

Many times these promotional groups welcome volunteers who will help with the program and in return they usually let the volunteers put up their own signs and talk with the customer about their own business. This can give you lots of exposure to interested customers in a program that would be too expensive for you to organize by yourself.

Check for a concrete promotion group in your area. The ready-mix suppliers and ICF distributors there might know about them. If not, call the staff at the Insulating Concrete Form Association or the Portland Cement Association to see if they can get you a contact. Then call the promotion group and see what programs they have planned and how you can get involved.

Listings

Don't forget those listings. When you join the Insulating Concrete Form Association you get listed on their web site as an ICF builder. You also get listed in the ICFA's annual directory, which is mailed to over 38,000 people. Both of these listings are free with membership.

The independent site *www.icfweb.com* has a directory you can sign up for, too. *Icfweb.com* and Permanent Buildings and Foundations magazine publish paper directories that you can sign up for as well. Get information from their web sites.

Figure 14-2 Sign to explain the benefits of ICFs and give contact information for the contractor.

All of these things reach buyers who can look under their state and find local contractors. You begin to realize the value of this easy way to get your name out when you start getting calls out of the blue from prospective buyers.

Signs and Street Traffic

A lot of contractors put out signs at their buildings under construction (see Fig. 14-2). But this is even a bigger opportunity with ICFs. Since ICFs are unfamiliar, they often attract a crowd. And people pay attention. According to Albert, an ICF contractor in Oklahoma:

> I've built a lot of stick houses and you get a few people who stop and look. But I built my first one with foam forms and man, people gawk and they come up and ask questions and want to know what it is and why you'd want to build with it and how it works and everything else. Some of them came and chatted for half an hour.

Put up a big sign that points out this is a new, better form of construction. And put your name and contact information on it in big, bold letters. This type of promotion works to get leads for you 24 h per day.

Word-of-Mouth

Don't forget the greatest advertisement of all, word-of-mouth. Satisfied customers are like gold. They tell their friends how great the building is and how great you are. Some will be references for you, and there's no one to put a prospective buyer's mind at ease better than someone else who has been in the same situation.

So try to develop a relationship, especially with your first few customers. Try to know them a bit. Bend over backward to make sure you do a good job for them. Follow up afterward and make sure they like the building. Ask if you can give their names out to new customers. If you have gold, polish it.

Be Persistent

Al, a contractor in Iowa, said:

> I'm more and more amazed at the times somebody calls me up from my phone number on a brochure I handed out two years ago and wants to talk about building him a house.

A lot of the marketing you do will pay off over the long term. And it accumulates.

Many people who plan to build a house don't actually do it for a year or two or three. They start doing their research, but they don't have definite plans. If you keep handing out the literature with your name on it and making the presentations, your information stays with them to tap when they do make their move.

The situation with architects is similar. According to David, an architect in Illinois:

> With architects, they don't usually get a presentation and say, "Wow, I gotta use that on this building I'm doing next month." The near-term buildings are already designed and committed. But some of them say, "Wow, I really coulda used that on this building I did last year." They file it away and when some project comes along that the product suits they consider using it.

So the information sits with them, too. Calls from either group can just keep rolling in over the years. You just need to keep putting your name out so that more and more of them have it.

All of your efforts build on each other. After a few years, the results from all the promotional work just grow and grow.

Be Observant

Wayne, an ICF contractor in Ontario, says:

> We built some ICF houses and just never got much interest. Then we got a job where the owners asked for a concrete floor, too, and that one got to be on display for a few weeks before they moved in. It was incredible. It was like turning on a light switch.

People really loved the idea of this house with both walls and floors built all out of concrete. Now that's all we build and we have a steady business.

Wayne's experience is *not* typical. But that's the point. In different places different things are popular. In Wayne's case the addition of the concrete floor sparked something with customers. In some areas a lot of middle-income buyers are determined to buy an ICF house, even though in most places the buyers come more from the high-end market.

So remember to be observant. Listen to buyers and watch how they react. They may not always want exactly what you thought they would, but this doesn't have to be a problem. You may find a whole new opportunity selling a different style of building or selling to a different type of buyer than you originally planned.

Overview

Certain things crop up again and again in discussions with ICF contractors and distributors. They often start out with "The one thing I wish I'd known when I got started with ICFs is…" or "My biggest piece of advice to someone who is just getting into ICFs now is…"

These are not the details. You'll learn the tricks of construction later. These are the bigger things you need to think about; the things that can affect a large part of your business and your operations. They are the big issues you have to keep in mind because they are really important and have a major influence on your success. And if you don't force yourself to bear them in mind, all the little details can push them out.

You have heard some of them before, earlier in this book. And that is fine. Read them again. They bear repeating.

Expect to Learn

Lyle, a distributor of ICF products in Alberta, says that the single most important thing he has to tell the new contractor is as follows:

> Expect that there will be a learning curve and he won't develop efficiencies on his first job. But as he does 3 jobs, the efficiencies will kick in. We had a few contractors who tried an ICF foundation or something and it took longer and cost 30 percent more than the usual way of building and they said "So we're not going to do any more." But that's just when they start getting real improvement, and they're wasting it. This also applies to the subtrades.
>
> The only solution is identifying the problem. Be aware of this and don't let it sidetrack you. You'll get efficient and your subs will learn to work with it. Some of the subs can even get faster than they are with conventional construction and end up liking ICFs better.

Many contractors talk about how important their frame of mind was to their first few jobs. If they were curious and eager to learn about ICFs the first project did not disappoint. How much they learn and how much they improve the first time out is exciting. But if they prepared poorly and assumed that they would make a killing on the first job, they could get frustrated. It's like the kid who joins Little League. He can make the most of it and work his way to being a good player, or he can get mad and quit because he wasn't the star the first day.

Try to Start Small

A good way to make the first project as rewarding as possible is to go out and get one that is a bit simple and straightforward. According to Mickey, an ICF contractor in California:

> Looking back, I'm really glad our first job was a little storage building. The inside had to be 70 degrees all the time, and the owner figured the foam would work out well. But the thing was it was hard to screw up. It was a box. One door and a roof. It was so simple I could've run almost double the labor I estimated and still not lost any money. But really we were pretty close to right. It's hard to screw up when you do those straight runs.

Different people say the same thing in different words. You want to take it easy on the first project or two because that's where you're going down the learning curve. You are more likely to make mistakes on your first project because it's all new. If it is simple, there are fewer mistakes to make, and the ones that do occur will have minor consequences. For your second project you can try something a little tougher. You know the basics, so you won't be trying as many new things all at once.

In areas with basements, they are a favorite first project. They usually have few corners and openings. In most cases everything is rectangular.

Allow a Little Extra Time

Another way to approach your early projects is to allow yourself plenty of time to think things over. Alan, a contractor in Massachusetts, says:

> The day before our first pour, I told the GC we were going to spend a day going fishing. He looked at me funny, but that's how you think of all those little things that you miss when you're rushing to get the walls up.

Randy, a Georgia ICF contractor, puts it this way:

> Make sure that you allow enough time for your first jobs. If you have decided to do something new don't make the mistake to rush through the first installations. Take your time to learn it the proper way, because the way that you learn things in the beginning is the way that you going to build.

And Joe, an ICF contractor in New Mexico, says:

> Don't rush the pour. Make sure everything is sealed and you are really ready when you call for the readymix. On one job I had gone in the morning and this guy had

already called the truck. But on inspection, he still had lots of work to get to the pour. So the truck was sitting there for a while, and that was a lot of money. Once you get the boom truck and concrete there, your clock is ticking, and you don't want to rush to do the actual pour either. Retrofitting with concrete, that's really expensive, to alter a door jamb, etc.

And finally, Cooper, a former contractor and current distributor in Minnesota, says simply:

Almost every mistake we see made in ICF work would have been avoided if the contractor had really gone through our pre-pour checklist point by point, like we tell them.

Don't Bother Trying to Find the "Perfect" Product

Danny, an ICF contractor in Minnesota, says:

Almost all the forms give you about the same final wall now. There are some differences in them that can make it easier for the installer to get them all installed and filled. But a lot of that is a matter of taste of what each guy likes to work with best. That's not the key thing that will make you or break you in this business.

Experienced contractors tend to agree that you need to spend more energy learning installation, improving operations in the field, and promoting yourself than you need to spend on trying to decide which is the "perfect" product. It pays to ask yourself which supplier will help you most to do these other things, rather than ask which product has the cleverest design.

Get Training and Get Help

Ian, a contractor and trainer in Texas, says:

We run a two-day course. On the second day we stack a little safe room-type structure and at the end we even pour it full of concrete and everybody gets to hold the hose. That gives them *familiarity* with everything. They can't really do everything they'll have to do in the field but they've seen it once and they understand the ideas. Then we invite them all to work on our crew for a week in Texas, and we pay them and everything. And in that week we can get through stacking all the walls for a story of a building or something and pouring them and they really get to do it all. *Then* they're usually in pretty good shape to go and actually build something.

The two critical elements in learning to build with ICFs are learning the basics before the first project, and having experienced help in the field.

The best way to learn the basics is to go through a training course. Everyone means to get some training, but the press of work prevents a lot of them from doing it. They almost always regret skipping the training, sooner or later. At the time they worry that they are behind and they have to get the job done. So they dive right in. But later they see that they would have worked much more efficiently and made fewer mistakes if they had gotten the instruction up-front. If you absolutely can't attend a formal course, watch instructional tapes or read the manual cover to cover.

And try *definitely* not to do your first project without experienced help. If you can arrange to work on someone else's project to cut your teeth, fine. If not, get your local ICF rep or an experienced contractor to come to the site to help out.

Get a Top-Notch Bracing/Scaffolding System

Cooper, a distributor and former contractor in Minnesota, says:

> The bracing is a critical component of the job. We know this because in the first few years of the business there was no manufactured bracing and we were always fussing and fretting to make things right. The manufactured systems available now are much, much better—work goes faster and the walls come out much better.

Contractor after contractor says that the bracing/scaffolding system is the one most important piece of equipment they have. The biggest quality issue with ICFs is getting the walls plumb, level, and square. These are things that have to be controlled quickly during the pour, which is precisely what a sturdy, adjustable bracing system is designed to do.

As Vay, an ICF contractor in Nevada, puts it:

> The biggest thing you have to tell newcomers is "Get good scaffolding and bracing!" Instead of just using wood, get the adjustable metal braces, because there are always the small after- pour adjustments, and those took a lot longer and were less precise with wood bracing.

So don't skimp. A good system will pay for itself many times over.

Work With People Who Want to Do This

According to Stanley, a general contractor in Texas who focuses on ICFs:

> I had to do some cajoling to get local framers to accept this. I picked my original crews according to their willingness. I have a couple now who do the work. You have to go through a few buildings with them and then you've established a norm. I don't do competitive bids—I ask a price and we negotiate. I have my own estimate calculated in advance. The best sub I have actually likes the system a lot.

Mark, an ICF homebuilder in Illinois, says:

> You want buy-in from everyone on the job. Not only the form installers, but the carpenters, electrician, plumber, and the HVAC contractor, who has to size things correctly.

It's plenty of work to learn a new building system. Try not to compound things by also working with skeptics any more than you have to. This even applies to what customers you work for, and what architects and engineers and building officials you work with.

Of course you don't always get a choice with these people. And you wouldn't want to hire an incompetent electrician because he's all excited about working with ICFs and the best electrician is lukewarm. But when you have some choice and when other things are more equal, it is helpful to favor the more enthusiastic people.

You can't tell them how to do every aspect of their job—they'll have to figure some things out. The enthusiastic person will make the effort to figure it out and do it well.

When you're stuck with a sub or an inspector or a designer who is skeptical about ICFs, it may be worthwhile to try to get the person more positive. You can point out the benefits. You ask a few questions to find out what the person *does* want and explain where ICFs will help achieve some of those things. With contractors you walk them through how the job is done. If possible, you get them in touch with someone else who was in the same position (did the wiring, inspected a building, designed an ICF building, or whatever) and felt good about the experience so that they can talk about it as colleague to colleague.

It doesn't always work out badly to push someone into ICFs. In fact, more often than not people see the value of the system by the time they finish the first project. But do your best not to settle for working with a lot of skeptics. Just think back on your last vacation where the kids weren't interested, and you have some idea of what it's like.

Estimation

Kurt, an ICF contractor in Utah, says:

> The thing I'd recommend that new contractors pay most attention to is estimating. Estimation is the single most difficult thing you have to learn. Your ICF company helps, but they can't provide detailed numbers because everything is different in different parts of the country. I figured out how to estimate, but it took 3–4 jobs. I didn't lose money, but it just takes some experience to get it really precise.

It's pretty clear that accurate estimating is important, because if you do it wrong it's likely to affect how much money you make. But, as Kurt says, the hard part is knowing exactly how to do it. Some ICF suppliers have good aids and their distributors may have lots of experience to share with you. That gives you a leg up.

But you get really accurate only after you've done a few. To help yourself, keep records of everything you can. Keep the blueprints and the list of everything you spent. Then on a rainy day you can list all the expenses for each project. This should begin to make it clear how costs vary. How much did cost seem to go up when the floor plan got more chopped up? What did it save when you got to the architect and engineer early and showed them how to make things simpler?

Market a Silk Purse as a Silk Purse

Steve, a longtime ICF homebuilder in Missouri, says:

> The biggest thing I wish I'd realized when I got started is that there *is* a market for better. We were in the market to be competitive. But that was a mistake. People buy a Rolex or Lexus because they perceive it as better. If I'd done a better job of marketing the "better" I wouldn't have had to work as hard to be competitive. I had a better product and *I* didn't even understand it, so I didn't do a good job of explaining it to customers.

What we found in my particular market was that the hospital had young doctors and anesthesiologists coming to work and they worked at night and had to sleep in the daytime. We had great sound properties with the ICF walls—they could sleep all day. I didn't even realize that until one of these guys told me, "Man, I would have paid ten thousand more if I'd known that." That's the hardest thing to put your finger on, but that's why a person buys an automobile—it's the comfort they feel.

The first advantage of a builder marketing his houses as better is that he will *immediately* increase his profits because he's no longer competing with the guy who's slinging lumber next door. He has the quality. The framing industry is so preoccupied with cheap and cost per square foot and throwing it up quick that they have just quit competing on quality.

Lon, an ICF contractor in Oregon, vouches for this approach:

> You're selling a Cadillac house, and you're not charging that much more. I don't have any trouble selling this to homeowners. I just have to explain it to them for ten minutes.

The hard part for a new ICF contractor appears to be breaking old habits. When the conversation turns to why he is charging more than someone else, the correct response is probably not to cut the price. More likely, it is to explain what the buyer is getting. Instead of saying "here is what you get", it may mean listening to what the customer wants in a building and finding and explaining how ICFs can provide some of those things while other systems cannot. It may even mean looking for customers in groups of buyers that will appreciate the benefits, instead of simply taking low-paying jobs from any source you can, then running ragged to fill a lot of low-profit orders.

If you find yourself listening to buyers who complain about the cost, consider the words of Joe, an ICF subcontractor in Pennsylvania who does mostly commercial work:

> [One of our big customers] complains that they can't justify the product. But they've done it on several projects. They've even built and leased office buildings for themselves. The speed we could do it with was the determining factor. They had us quote on an office building that was all gables and corners. Our quote jumped from our normal $36 per block to $50 per block. They complained and asked why it was so much and we explained. They said they couldn't see how it was justified. Then they hired us. That's fine with me. I can listen all day to why they can't pay us that much as long as they keep paying it.

Go Big or Go Home

The advice from Ken, a former ICF contractor and current distributor, is:

> Go big or go home. If you want to do ICFs, totally immerse yourself in them and make them the focus of your business. You can't dabble in it over the long term. You need to become the go-to guy, the specialist who can do the job.
>
> I decided I'd become the go-to guy and developed a contractor network. If you just immerse yourself in it your jobs will go better and become faster if you're in it all

the time. If you don't do that, you'll be outcompeted by others who do. At 2–3 jobs per year consider it a hobby. You won't be particularly successful at it.

Will, an ICF distributor in Georgia, says much the same thing in another situation. He notes that many ICF contractors got their start in the business doing double duty. They not only built with the forms, they sold them. They signed on to act as distributors for one of the ICF manufacturers, and sold forms to other contractors while they built things with those forms themselves. But Will says:

> I think ICF guys have to decide what business they are in. If you're an ICF builder, you need to be the best builder you can be. If you're an ICF distributor, you need to be the best distributor you can be. If you try to do both, you hit a plateau. You can't grow your distribution business past a certain point because you're selling to your competitors and they'll resist dealing with you too much. And it's tricky to run your construction business because people think you're tied to one product because that's the one you sell.
>
> Initially in this business the only way to make money distributing the product was to provide the whole construction package—the product and the installation. And it's tempting to do installation and get 80,000 dollars for a project instead of just selling the forms and getting 15,000 dollars. But if you do you hit that plateau in distribution. I haven't contracted a building in 4 years, and now my distribution business is growing steadily. And if you are a contractor and you get tempted to sell the forms, too, then you'll hit a plateau in that business. You have to make up your mind what you want to do and then do it right.

You can always start out building a couple of ICF buildings to see if they're for you. And you can try doing different jobs in the ICF industry. But before long you would be wise to take a hard look at your activities and see where your strengths and interests lie, then focus there. Do I still prefer frame? If so, go back and be happy. Do I prefer ICFs? Then commit to them fully. Do I want to be in another end of the business? Then maybe you shouldn't try to be an occasional builder. If you try to do several things it is likely that you will only be fairly good at each one of them. You stand to lose business in every thing you do to specialists who do it better.

Marketing Flexibility

Clif, a contractor in Arkansas, says:

> When we started I told everybody an ICF house was like living in "one big cooler" and I went after some of the really big houses. But a guy came up to me and said "You know, I could use something like that for my crops." I didn't want to do a farm building, I wanted to do houses. But he kept coming after me and I hit a spot where I had to have more work, so I did it. It was kind of a pain because we had to figure out what you have to cover the walls with for an agricultural building and things like that and I didn't have any experience at that. But otherwise it went okay and I guess the building worked well because he told his friends and after that I learned how to get the word out to farmers. It turns out it can be really good for storing fruits and vegetables and some things that you have to keep cool. Now the (agricultural)

buildings are almost half my business and they can be strong sometimes when other stuff is slow.

The market is changing and you can benefit by keeping your eyes open. What once was a product that sold mostly to high-end homes is spreading to other segments of the housing market and finding good markets in commercial construction, too.

This is one area where you don't always have to decide between one option and another. You may be able to do a good job building more than one type of project. It can even be useful for keeping your volume of work up, and for evening out some of the ups and downs that come in different segments of the market.

Pay attention to how people in your area react to ICFs—who is interested, who wants to build with them. You don't want to spread yourself too thin. Doing a new type of building every time out is inefficient. But if you hit on a strong demand for one type of building, it might give you a good opportunity. It might be better than the market you're in now, or it might be profitable to focus on two markets instead of one.

Because ICFs are finding more and more uses over time, no one can tell you every place to look for business. It's up to you to be alert.

Remember that Commercial is Different

Peter, a longtime ICF contractor and distributor in Florida, says:

The crews in commercial construction have more experience with concrete, but commercial is a lot more planned, too. In residential construction, if a home is designed as a CMU or poured basement, with wood frame on top, conversion to ICF is simpler and much of that can be done during the building process, if the code officials allow it. Many times there is no need for a structural engineer, as the architect or the prescriptive code method may cover all details and steel reinforcement.

In commercial construction, civil, structural and mechanical engineers are usually involved in the design, together with the architects. If these professionals never have worked with ICFs before, it is very important to educate them to avoid design details to be specified that may lead to building problems in the field.

Education of supervisory personnel and other trades is also critical, especially if they have never worked with ICFs before. Explain the dimensions of the form, which affect placements of plumbing, electrical, mechanical, low voltage, sprinklers and other utilities. Can they be placed in the foam or do they need to be sleeved and poured in? What preparations need to be made to achieve fire ratings for walls and penetrations?

When building, involve the architects and engineers in the process; they usually are responsible for all the inspections. The better they understand the system, means and methods, the smoother the project will proceed. Frequent "in progress inspections" may be helpful, especially at the beginning, until every body understands each others' concerns, requirements and limitations.

Commercial projects are usually on a very strict schedule, most times with a liquidated damages clause for the general contractor built in. On-time performance is critical for the ICF installer/builder and all other subs. It is important to size and organize the crew well to be able to perform to the schedule. Forms, materials and bracing must be delivered on time, and delays due to weather should be planned in, with alternates available. 24-7 work may be required to avoid delays.

As there is much more interaction with large crews of other trades, scheduling of the work and job safety become important issues.

Building the walls is the least difficult item—straight, plumb and square and gravity are the same as in residential construction. Commercial walls may be taller, there may be more concrete pressure, and certainly more volume. The building may be larger, it may need multiple pumps on pour day, and multiple crews working on opposite ends. The ICF builder may need to step back to organize the show, to avoid problems, and not hold the hose himself, as is usually the case on small residential jobs.

Most of this is standard operating procedure with any commercial construction. However, it's especially important for ICF contractors to hear it because they frequently make the jump from residential to commercial. Most ICF contractors are in homebuilding and they switched from frame. But after a while they feel the pull to do commercial projects. ICFs are natural for many types of commercial construction—they can be very competitive and often the contractor can make big margins. There are, also, not many ICF crews in many areas yet, so the owner or GC can nudge the local ICF homebuilder to try to do the work for him. All of this is fine so long as the contractor pays attention to the things that are different in commercial construction. It is not just building a "big house".

Beware of Being Too Slow

Ray, an ICF contractor in North Dakota says:

> The biggest thing I wish I'd done differently is I just wish I had done it (started with ICFs) sooner! I first learned about ICFs more than five years ago, but was too chicken to try them out for a while. I had been building with stick framing, and it seemed like too much new stuff to learn to build with ICFs. Then when I finally got in touch with an ICF manufacturer, they sent out two techies to help train my crew, and they came back for the first job and logged a lot of hours with the crew. My first house went nearly flawlessly. It's been great.

Changing over to a new building system is nothing to do lightly. You need to do your homework and make sure it makes sense. But there is danger also in being slow. Other people are switching to ICFs in new towns every day. You may be skittish about being first, but it's potentially a lot worse to be last. If you have doubts, remember that you don't have to make a lifetime commitment in your first year. You can, like Ray, do one project and see how it goes.

Have Some Fun

Why have fun with this? As Scott, an ICF contractor in Arizona, says:

> Why not? When you get down to it this is pretty cool stuff and it's fun to learn something new and it's fun to build a better house than everybody else and talk about how it works.

16

The Future of ICF Construction

Overview

ICFs are young by construction industry standards. And when a product is still young, it is still changing. People revise the design to improve it. They refine the way they work with it. And they adjust the ways they support, promote, and regulate it. So it will be with ICFs.

A lot of the change has already occurred, and ICFs are not going to transform radically in the next 5 years. They will still look pretty much the way they do now. But there will definitely be differences to look forward to.

Growth

ICF construction is currently growing at the rate of about 30 percent per year. That can't go on forever, of course, but it can go on for several more years. Under 5 percent of all houses are built with above-grade walls constructed of ICFs. The figure is even lower for commercial construction. So there is plenty of room to keep growing at the current rate for 5 to 10 years.

Over that period you should expect to see ICF construction come to the places where it is now only a rumor. If you live in an area where nothing has ever been built with ICFs, more than likely someone will introduce them before long. If you live in an area where they are used here and there, you can expect them to become a regular construction method along with wood frame, concrete block, metal buildings, and the like.

They will continue to grow not so much because people are discovering them (although that is a big part of it). They will continue to grow because they are *where the market is going*. The benefits of ICFs are precisely what more and more buyers and the building codes are requiring. Every year people's expectations increase about how tightly controlled the indoor environment of their buildings will be. Every year the code requirements for the energy efficiency and disaster resistance of buildings ratchet up a little. And as these requirements get higher, more and more owners and architects are discovering that ICFs help them meet these more stringent demands. Systems like wood frame can change to meet them, but only by adding extra parts and materials that steadily increase the time, cost, and difficulty of construction.

Take an example of something like this that happened before. Loren, a general contractor in Florida, said:

> When I started construction in the '80s, maybe seventy percent of the houses in Florida were frame, and the rest were concrete—block, mostly. It was just cheaper to build with frame by maybe two thousand dollars a house. But then Hurricane Andrew hit, and everyone saw all those pictures in the newspaper of houses picked apart and laid down in piles of sticks on the ground. Well, the cities reacted by getting more strict with their code enforcement. You had to prove to them that your house would meet the wind requirement and in a lot of areas every house had to be engineered. When the plans came back the frame houses all had to have a lot of steel strapping and connectors added. The engineers were telling us that the frame designs everyone was using weren't really strong enough to resist the high winds. I guess we had just slipped into building lighter and lighter and the inspectors hadn't picked it up. Well darned if the cost of building with frame didn't go up by about two thousand dollars a house. Meanwhile, the plans for the block houses came back with hardly any changes at all. Concrete block walls pretty much already met the wind requirements without adding anything. So all of a sudden the costs of a frame house were somewhere around the same as with concrete. So which do you build with? If the cost of construction is the same with both, it's a wash, right? Wrong. If you're a homebuyer and the price is about the same and you saw all those pictures of Andrew, you buy concrete. And that's what they started doing. So builders could hardly sell a frame house in a lot of areas. Now sales are pretty much the other way around. There are maybe seventy percent of the new houses that are concrete.

So, what happens in the future when the codes and the people start asking for disaster resistance and more energy efficiency and tighter construction and mold resistance and maybe a quieter interior to boot? The builders and architects find that they are struggling to achieve all these things with many other building systems, but they pretty much come as standard equipment with ICF construction. Sales numbers in recent years support the idea that this is where the market is going, and will probably continue to go in the future.

Technology

There are changes and improvements in ICF design and installation that happen every year. Many of these you don't even see. These technological improvements are hard to predict. But we can already see where companies, contractors, and inventors are spending the most time and money. There are a few changes that a lot of people are working on, and it's a good bet that these will be some of the more visible advances in technology in the next 5 to 10 years.

One area ripe for change is **mix design**—the recipe for the concrete used in ICFs. Contractors are constantly looking for a concrete mix that will flow well through the pump and into the forms, that will keep the pressure on the forms low, and that won't cost too much. If they can improve the flow and reduce the pressure, they can fill the forms faster, yet keep air pockets out and avoid getting the forms out of alignment. They have thrown all sorts of combinations of ingredients at the job, and many have their own personal mix designs using standard materials.

But now the suppliers of admixtures are coming up with new ingredients and new ways to combine them. More and more suppliers are devoting a lot of effort to figuring out how they can best combine them to make a mix ideal for ICF construction.

The latest excitement in the field is **self-compacting concrete**. This is concrete with a combination of admixtures that makes the mix flow almost like honey—faster but steady, without the sudden rushes you can see from ordinary high-slump concrete. It's too early to tell whether a self-compacting concrete will invade ICF construction. But it's clear that now the suppliers have the ingredients to put together a better mix for the job—something that lets the contractor get off the concrete pour much faster and leave a great set of walls behind him when he does. It's just a matter of time before the details get worked out.

Another area that may be changing is the steel reinforcement. Adding steel rebar is an inexpensive and pretty quick way to increase the strength of the ICF wall up to almost anything you want. But in the push for better and faster, suppliers are now experimenting with **fiber-reinforced concrete**. This is concrete with steel and plastic fibers added in before it's placed into the forms. It can add considerable bending strength to the hardened material. This adds to the cost of the mix, and you still need rebar. But people are experimenting now with using more fibers and less bar. There is less labor in putting the bar in position, and the wall cavity is more open to make it easier for the concrete to flow in during the pour.

HVAC

Getting good HVAC work on their buildings has long been a big sore point for ICF builders. Nervous heating and cooling contractors are often skeptical of the claims made for ICF walls, so they install the same equipment they would in a conventional building. But this is usually inappropriate. Fortunately, information is getting into the field that should correct the problem in coming years.

Sizing HVAC equipment

HVAC contractors tend to doubt that ICFs save as much energy as people claim. They don't want to be blamed if the equipment is too small to keep the indoor temperature at the set point, so they put in the same size of equipment they normally would. The money that should be saved on smaller equipment is not. Just as bad, the oversized equipment short cycles—it only runs for brief periods, during which it blasts in the hot or cold air and quickly brings the indoor temperature back to the set point. This is inefficient, hard on the equipment, and it doesn't give air conditioning enough time to take the moisture out of the air in humid climates.

Fortunately, help is on the way. The Portland Cement Association and the U.S. Department of Housing and Urban Development have sponsored engineers at Construction Technology Laboratories to come up with methods to do accurate sizing of HVAC equipment for small concrete buildings. At this moment, the work is nearly done and should be released to the public shortly. The estimation should be available in the form of a computer spreadsheet that is inexpensive and easy to use. It is the result of careful engineering work and comes from an authoritative source, so it should help ICF builders make their HVAC subs comfortable with smaller equipment.

As the new sizing methods find their way into the hands of the HVAC contractors, and as they get discussed and recommended at trade shows and in the press and by different experts, the contractors will gradually learn to trust them and use them. And even if many of them never use them, they will still probably have broad influence. Other HVAC sizing software and manuals will probably adopt some of the new methods. More than likely some of the key calculations will be reduced to simple rules of thumb that the contractors use just as they now have simple rules for sizing equipment in frame buildings. With the message getting to contractors through all these channels, we can expect that over time the sizing of the equipment will improve, and the cost of HVAC systems in ICF buildings will come down.

Fresh air exchange

The other area of uncertainty in HVAC design has been whether and how much fresh air to bring into a building. Fresh air intake is not required in most codes, but as buildings get tighter and tighter, more people are recommending it and more contractors of all types are incorporating it. ICF buildings are a step tighter than most conventional buildings because of their walls, even without taking steps to seal other parts of the building shell. Currently about half of ICF buildings are constructed with some form of fresh air intake.

In the next few years it is likely that mechanical fresh air intake will find its way into nearly all buildings. Sentiment is growing for it to be a requirement. The American Society of Heating, Refrigeration, and Air Conditioning Engineers (ASHRAE) has voted to adopt so-called Standard 62.2, "Ventilation and Acceptable Indoor Air Quality in Low-Rise Residential Buildings." This Standard contains rules that would require regular fresh air ventilation in nearly all homes, with specifics about how this could be accomplished.

An ASHRAE standard is not law, but once it is established it is likely to get picked up eventually by the codes and by HVAC contractors. Standard 62.2 is not yet final and there are several groups still challenging it. However, with a lot of support behind this kind of change, mechanical fresh air ventilation in homes and other small buildings is likely to become standard over the coming years. This should go a long way to taking the uncertainty out of the decision of whether and how to ventilate a new building.

Related and Accessory Products

Already showing up are whole new families of accessory products just for ICFs. Some construction products companies, like Wind-Lock, have set up whole divisions or product lines geared just for ICFs. Even some whole new companies, like ICF Accessories, have been formed just to design and sell new equipment and accessories just for ICF construction. They offer tools, connectors, scaffolding, waterproofing materials and the like that were designed from scratch just for ICFs.

One good example of a product designed just for ICFs is the ICF Ledger Connector, sold by Simpson Strong-Tie. This special connector enables the contractor to attach floor ledgers to ICF walls relatively quickly, cleanly, and economically.

Another good example is the range of new waterproofing products that have been designed or modified for ICF foundations by established companies like Polyguard, Meadows, and Platon, as well as by new companies like Advanced ICF Products and Aquaseal USA. Polyguard's new waterproofing membrane also is code-approved as a termite barrier, so it doubles as termite protection in the high-risk areas where the codes require termite protection over the foam below grade.

There are also several lines of scaffolding developed specifically for use with ICFs, too, from such established and new companies as ReechCraft, W.A.S.S., and Plumwall.

The improvements in accessories and equipment here are coming fast. V-Buck invented a popular line of plastic buck material to use in place of lumber. Schwing now offers a line of reconditioned boom pumps to concrete pumping companies that are geared for ICF work and carry a much reduced price. And there's more—the list goes on and on.

The growth and refinement of accessory products will continue. They will help build straighter walls faster, attach floors and roofs and fixtures more quickly and securely at a lower cost, and cut and shape the walls more precisely in less time and with less money.

Familiarity

More and more people are getting familiar with ICFs, making it easier and easier to sell them, find subs, and get the buildings approved. According to Clint, an ICF contractor in Nevada:

> It used to be at home shows everybody asked "What are these things?" and "How do they work?" and "Why should I want them?" Now it's more like "Are they available in my area?" and "Who can build for me with them?" They already know about the forms from shows like "This Old House".

Contractors across most of North America are reporting that they have to explain ICFs less and less to customers. And more and more buyers are doing their initial homework on their own and picking ICFs before they ever talk to a contractor. Phil, a distributor in Massachusetts, explained:

The City of Boston has to rebuild the "area ways". Those are the little tunnels that go under sidewalks and into buildings like stores to get merchandise in. Their engineering department did a lot of research and they decided that ICFs would be a very cost effective way to do it. They have a lot of changes in height and cutouts and stuff—it's easier to make those things just by cutting some foam. They called on us and asked for specs and what kind of pricing we'd give and things like that. They didn't need to be convinced about the product at all.

This is bound to continue. In the future it will be necessary to do less and less explaining about ICFs to buyers, and instead you will be called on to explain the things every contractor is asked—what experience you have, when you're available, what references you can give, and so on.

Finding crews and subs will also get progressively easier. Subcontractors—mostly framers—are getting into ICF wall construction at a rapid clip. The United Brotherhood of Carpenters alone has over half-a-million members, and the organization is offering comprehensive ICF training to almost all of them. They have the capacity to train several thousand carpenters per year. Soon every area will have at least one electrician, plumber, wallboard crew, and HVAC contractor who has worked on ICF buildings before.

The building departments are getting familiar, too, and many are getting comfortable with ICF construction. This is largely because they are seeing more and more ICF buildings, like everyone else. But it's also that ICFs are getting covered more and more thoroughly in the building codes.

ICFs are now thoroughly covered in the International Building Code. It now looks likely the International Building Code will replace the three U.S. regional codes (Basic from BOCA, Standard or "Southern" from SBCCI, and Uniform from ICBO) and become the model code adopted by the majority of state and local building departments around the country. As this happens it will give the local building officials increasing confidence in ICFs, and answer most of their questions so that the contractor doesn't have to.

In Canada the National Building Code does not yet include sections specifically on ICFs, but that should within 2 years. The 2005 version of the code is mostly drafted already, and ICFs get pretty thorough coverage in it. Like the International code in the United States, the new version of the NBC should get adopted by more and more Canadian jurisdictions over time.

Production Building

ICFs are now going into some really big housing developments. And as you would expect, the contractors doing the work are streamlining things to make construction more like an assembly line.

At the time this is being written, work has just begun on The Villages at Rio Del Sol in Cathedral City, California. It is slated to have 160 homes constructed out of ICFs. According to the supplier:

They plan is to build the [first] six homes in three to four weeks and then roll into a four-house-per-week schedule, though that may increase to six homes per week.

In brief, the system calls for marking all interior and exterior walls throughout the site at once. These lines will be coated to protect them. The crews will then put up the [ICF] walls, bracing them on the outside. Another crew will put together movable scaffolding, which will be used to fill the concrete within the walls. Once the walls are set, workers move the scaffolding on to the next house. The builder's crew can begin framing the interior walls the day after placing the concrete. This process will save one-and-one-half to two days work per house.

The development of production building is predictable. It happens whenever a building system gets used in large developments. In addition to Rio Del Sol, there are plans on the boards for large ICF developments in Texas, Minnesota, and probably a dozen other places.

As the popularity of ICFs grows, developers will feature them to get a marketing edge. They will advertise that their houses are superior because they are constructed with ICFs. And the need to construct a large number of ICF houses in a short period of time will lead them to develop streamlined construction systems. As with other building systems, this will probably end up consisting of several specialized crews, each one doing a part of the job and rotating from site to site.

Prefabrication

Prefabrication may be a big part of the push to find more and more efficient ways to install ICFs. There are several contractors around North America who already specialize in it. They build their wall panels in a plant or a yard, gluing the forms together. Then they truck them to the site. They are light enough for two to three workers to lift them into place.

Prefab panels have carved out a niche in frame construction, and they are likely to carve out at least as big a niche in ICF construction. One factor that limits the use of prefab panels is that small building construction always has a lot of change orders. The plant can build all the panels perfectly and send them to the site, but when the customer changes to a different size of cabinet or moves a window, a lot of panels have to be ripped up and redone in the field anyway. And this kind of thing will not go away.

But ICFs have some advantage for prefabrication. One is lightness. Shipping and lifting ICF panels does not require heavy equipment or a lot of muscle power. So they are a bit more practical. Another advantage is that when changes are necessary it is fairly easy to cut and paste the foam. So all in all, it is likely that the next few years will see a healthy expansion of the business of prefabricating ICF wall panels.

Commercial Construction

It's a no-brainer that before long ICFs will grow faster in commercial construction than in residential. ICFs are a bigger share of the residential market and are only now getting used in commercial in a big way. So there's more room to grow in commercial.

On top of that, ICFs are ideally suited for a lot of commercial projects. The higher strength and durability requirements of commercial are the reasons concrete in various forms is already such a big part of the construction there. ICFs have those same important properties. And to boot, they are often *less* expensive than the other wall systems currently used in commercial construction. So watch for ICFs to go into that next office or hotel or industrial building in town. And then watch for them to go into more and more and more.

Distribution

Years ago when the ICF industry started, the local product distributors were mostly contractors who had used the product and decided to make a few bucks on the side selling it to their friends, too. But more and more, product distribution is going mainstream. The newest distributors are mostly already established suppliers of common construction products. They include ready-mix, masonry, and lumber supply companies. They tend to be larger, more professionally organized, and carry a lot more inventory.

This will probably become the norm over the next decade. ICFs will be handled like any other mainstream construction product.

There are even attempts to sell the forms through the big home centers, almost like they were 2 × 4s. It's probably a long time before that is very successful. ICFs are still a more technical, engineered product that the contractor needs a fair amount of support on. But it does show just where things might be heading in the long term.

Total Concrete Shells

A decade ago maybe eighty percent of the projects using ICFs had them in the basement only. Now perhaps eighty percent have them in all the exterior walls, basement and above-grade. And now a growing number of buildings use the new concrete deck systems to construct the exterior walls, the floor decks, and the roof all out of concrete.

The appeal of the "total concrete shell" is obvious. If you buy ICF walls for energy efficiency, disaster resistance, reduced sound transmission, durability, and indoor comfort, you can increase all those things even more by replacing the frame in the roof of the house with foam and concrete, too. The concrete floors can add to these benefits as well.

But in the past the cost and logistics of installing horizontal concrete in the air kept this sort of thing to a minimum. Now things are starting to shift. Better and better floor and roof systems are getting developed that are faster and promise to be more economical. At the same time, demand is shifting to favor the benefits that concrete shells offer, and people are gradually getting more money to pay for them. So the building with the "total concrete shell" has appeared, and is now becoming a regular type of project for some contractors. In all likelihood it will be a larger and larger share of ICF construction over time.

The development of the total concrete shell will require developments in building ventilation as well. With all walls and the roof built of ICFs, the building will have markedly lower air infiltration than it would with conventional construction. So probably these buildings will need more advanced ventilation systems as well. But these systems are already under development. ICF total shell buildings will just take advantage of the new ventilation technology.

Cost

It's a good bet that the cost of ICF walls will gradually drift downward. For one thing, that almost always happens with new products. For another, all the forces to drive it are in place.

Many of the inventions coming down the pike shave some pennies off the job. The ICF suppliers are putting in more factories around the country to cut shipping costs. They are streamlining designs and ramping up production, which gradually cuts manufacturing cost. The shift to mainstream distribution will cut the time and cost of getting orders to the buyer. Contractors are becoming better trained and more experienced. This moves them down their learning curves so that they can build more and more efficiently.

Exactly where costs will end up is impossible to say. But knowing this is not very important, either. Already ICFs are less expensive than the competitive walls systems in many commercial projects. In houses the cost may or may not ever equal that of basic frame construction, but it will probably come pretty close. And that is close enough, because basic frame construction is becoming less and less adequate for our wealthier, more demanding consumers and our stricter building codes.

Ratings and Awards Programs

There are several new industry and government programs under development that would give even more financial incentives or market recognition for energy-efficient and safer construction.

Energy efficiency

At the time of this writing, the U.S. Congress is considering passing a law that would give builders a tax credit of $2,000 for each house they construct that meets higher-than-code energy efficiency standards. This would almost immediately make ICF houses more cost-competitive with ordinary frame.

Energy-efficient mortgages (EEMs) should soon become more widespread and easier to apply for. EEMs allow buyers of energy-efficient homes to borrow more money so that they have more money to spend on the house, or they provide some other favorable loan terms. ICF houses usually qualify for EEMs but the workers at the mortgage companies that have EEMs on their books aren't always aware of them. So currently it can be difficult getting them to give you information or provide an EEM.

But there are some steps underway to make EEM applications more routine. The Federal National Mortgage Association ("Fannie Mae"), the federal organization that provides financing for a lot of the mortgages in the United States, has developed standard rules for EEMs on the mortgages it finances. Fannie Mae is now working on a program to explain the procedure better to the mortgage lenders. The lenders actually provide the mortgages to the public, so if they understand how things work better they will be better able to help the borrower apply for and get the EEM.

In addition, several of the ICF companies are working with mortgage lenders to get them to provide EEMs to buyers of houses built with the companies' forms. Their focus is to send their buyers to a reliable lender offering energy efficient mortgages and an easy application process. Check with your ICF supplier about this.

Safety

There is also a new safety-rating program that may qualify buildings for lower insurance rates and make them more marketable to safety-minded buyers. It's called "Fortified for Safer Living," and it is administered by the Institute for Business and Home Safety (*www.ibhs.org*). This institute is supported by the insurance industry to help reduce losses of life and property.

Under "Fortified for Safer Living," program officials check the plans and the construction of a building, and if it meets their requirements for things like resistance to fire and high winds, the building gets an official certification. IBHS plans to link this certification to the building's insurance, so that the building owners pay lower rates. The certification can also be a selling point— buyers see that the building has been inspected for an extra margin of safety and has been passed.

The *Fortified* program is just now getting underway, but the IBHS is aiming to get it widely recognized by insurers. The advantage to ICFs, of course, is that an ICF building would probably qualify for certification without as many additional measures as other forms of construction would require.

You

It's pretty clear now that ICF construction will continue to grow, continue to be refined, and continue to establish itself as a mainstream construction product. Your question for yourself is where you want to be in all of this. You can lead or you can follow, and there is value and honor in either one. But you will have to do one or the other. We hope this book has given you some useful tools to help you make the shift when the time is right.

Index